Modelling with AutoCAD 2004

Other titles from Bob McFarlane

Beginning AutoCAD   ISBN 0 340 58571 4

Progressing with AutoCAD   ISBN 0 340 60173 6

Introducing 3D AutoCAD   ISBN 0 340 61456 0

Solid Modelling with AutoCAD   ISBN 0 340 63204 6

Assignments in AutoCAD   ISBN 0 340 69181 6

Starting with AutoCAD LT   ISBN 0 340 62543 0

Advancing with AutoCAD LT   ISBN 0 340 64579 2

3D Draughting using AutoCAD   ISBN 0 340 67782 1

Beginning AutoCAD R13 for Windows   ISBN 0 340 64572 5

Advancing with AutoCAD R13 for Windows   ISBN 0 340 69187 5

Modelling with AutoCAD R13 for Windows   ISBN 0 340 69251 0

Using AutoLISP with AutoCAD   ISBN 0 340 72016 6

Beginning AutoCAD R14 for Windows NT and Windows 95   ISBN 0 340 72017 4

Advancing with AutoCAD R14 for Windows NT and Windows 95   ISBN 0 340 74053 1

Modelling with AutoCAD R14 for Windows NT and Windows 95   ISBN 0 340 73161 3

An Introduction to AEC 5.1 with AutoCAD R14   ISBN 0 340 74185 6

Modelling with AutoCAD 2004

Bob McFarlane
MSc, BSc, ARCST
CEng, FIED, RCADDes
MIMechE, MIEE, MIMgt, MBCS, MCSD, FRSA

Curriculum Manager, CAD and New Media, Motherwell College

ELSEVIER

AMSTERDAM • BOSTON • HEIDELBERG • LONDON • NEW YORK • OXFORD
PARIS • SAN DIEGO • SAN FRANCISCO • SINGAPORE • SYDNEY • TOKYO

Newnes is an imprint of Elsevier

Newnes

Newnes
An imprint of Elsevier
Linacre House, Jordan hill, Oxford OX2 8DP
200 Wheeler Road, Burlington, MA 01803

First published 2004

British Library Cataloguing in Publication Data
A catalogue record for this book is available from the British Library

Library of Congress Cataloguing in Publication Data
A catalogue record for this book is available from the Library of Congress

ISBN 0 7506 64339

For information on all Newnes publications
visit our website at http://books.elsevier.com

Typeset by Charon Tec Pvt. Ltd, Chennai, India
Printed and bounded in Great Britain

Contents

Preface

This book is intended for the AutoCAD 2004 user who wants to learn about modelling. My aim is to demonstrate how the user can create 3D wire-frame, surface and solid models with practical exercises backed up by user activities. The concept of how multiple viewports can be used to enhance drawing productivity will also be discussed in detail. The user will also be introduced to rendering.

The book will provide an invaluable aid to a wide variety of users, ranging from the capable to the competent. The book will assist students on any national course which requires 3D draughting and solid modelling, e.g. City and Guilds, BTEC and SQA as well as students at higher institutions. Users in industry will find the book useful as a reference and an 'inspiration'. The book will also prove useful to the Design/Technology departments in schools who are now becoming more involved in computer aided design.

Reader requirements

The following are the requirements I consider important for using the book:
a) the ability to draw with AutoCAD
b) the ability to use icons and toolbars
c) an understanding of how to use dialogue boxes
d) the ability to open and save drawings from/to a named folder
e) a knowledge of model/paper space would be an advantage, although this is not essential.

Using the book

The book is essentially a self-teaching package with the reader working interactively through exercises using information supplied. The various prompts and responses will be listed in order and icons and dialogue boxes will be included where appropriate.

The following points are important:
a) All drawing work should be saved to a named folder. The folder name is at your discretion but I will refer to it as **MODR2004**,
   e.g. open drawing MODR2004\ MODEL1 or similar.
b) Icons will not be displayed, as the user should know how to activate icons
c) Menu bar selection will be in bold type, e.g. **Draw-Surfaces-3D Face**
d) Keyboard entry will also be in bold type, e.g. **VPOINT**, **UCS**, etc.
e) Prompts will be in typewrite type, e.g. `First corner`
f) The symbol **<R>** will require the user to press the return/enter key.

Standard sheets/templates

During the various exercises, the user will be required to create several standard sheets/templates/prototype drawings. The user can:
a) use the given information to create the given standard sheets
b) use their own standard sheets
c) contact me by e-mail and I will send the appropriate sheets as attachments. My e-mail address is: BMCFARLA@MOTHERWELL.CO.UK

Note

All the exercises and activities have been completed using AutoCAD 2004. I have tried to correct any errors in the drawings and text, but if any error has occurred, I apologise for them and hope they do not spoil your learning experience. Modelling is an intriguing topic and should give you satisfaction and enjoyment.

Any comments you have about how to improve the material in the book would be greatly appreciated. I would also appreciate any new model ideas from readers, as it is becoming more and more difficult to think up new concepts.

To CIARA, our beautiful
grand-daughter

Chapter 1

The basic 3D standard sheet

To assist us with the models which will be created, a standard sheet (prototype drawing) will be made with layers, a text style, dimension styles, etc. This standard sheet will be saved as both a drawing file and a template file. It will be modified/added to as the chapters progress.

1   Begin AutoCAD 2004 and from the Startup dialogue box select the Use a Wizard option and:

prompt	Startup (Use a Wizard) dialogue box
respond	**pick Advanced Setup then OK**
prompt	Advanced Setup dialogue box
respond	select the following to the various steps:

    a) Step 1 Units: Decimal; Precision 0.00; Next>
    b) Step 2 Angle: Decimal Degrees; Precision 0.0; Next>
    c) Step 3 Angle Measure: East(0); Next>
    d) Step 4 Angle Direction: Counter-Clockwise(+); Next>
    e) Step 5 Area: Width 420 and Length 297 (i.e. A3)

then	pick Finish
and	a blank screen will be displayed

2   *Layers*

Menu bar with **Format-Layer** and make five new layers:

name	colour	linetype
MODEL	RED	continuous
TEXT	GREEN	continuous
DIM	MAGENTA	continuous
OBJECTS	BLUE	continuous
SECT	number: 96	continuous
0	white	continuous

NB: other layers will be added as required.

3   *Text style*

Menu bar with **Format-Text Style** and make a new text style:
Name: ST1
Font: romans.shx
Height: 0; Width factor: 1; Oblique angle: 0
Apply then Close the dialogue box

4   *Units*

Menu bar with **Format-Units** and:
Units: Decimal with Precision: 0.00
Angle: Decimal Degrees with Precision: 0.0
Units to scale drag-and-drop content: Millimeters

5   *Limits*

Menu bar with **Format-Drawing Limits** and:

prompt	Specify lower left corner and enter: **0,0 <R>**
prompt	Specify upper right corner and enter: **420,297 <R>**

6 *Drafting Settings*
Menu bar with **Tools-Drafting Settings** and use the tabs to set:
a) Snap: 5 and Grid: 10 – not generally used in 3D
b) Polar Tracking: off
c) Object Snap: off and all modes: clear
   Object Snap Tracking: off

7 *Dimension style*
Menu bar with **Dimension-Style** and:
prompt   Dimension Style Manager dialogue box
respond **pick New**
prompt   Create New Dimension Style dialogue box
respond 1. New Style Name: **3DSTD**
           2. Start With: ISO-25 (or similar)
           3. Use for: All dimensions
           4. pick Continue
prompt   New Dimension Style: 3DSTD dialogue box
respond **pick Lines and Arrows tab and alter:**
           1. Dimension Lines
              a) Baseline spacing: 10
           2. Extension Lines
              a) Extend beyond dim lines: 2.5
              b) Offset from origin: 2.5
           3. Arrowheads
              a) both Closed Filled
              b) Leader: Closed Filled
              c) Arrow size: 4
              d) Center Mark for Circles: None
then   **pick Text tab and alter:**
           1. Text Appearance
              a) Text Style: ST1
              b) Text Height: 5
           2. Text Placement
              a) Vertical: Above
              b) Horizontal: Centred
              c) Offset from dim line: 1.5
           3. Text Alignment
              a) ISO Standard
then   **pick Fit tab and alter:**
           1. Fit Options
              a) Either the text or the arrows active (black dot)
           2. Text Placement
              a) Beside the dimension line active
           3. Scale for Dimension Features
              a) Use overall scale of: 1
           4. Fine tuning: both inactive i.e. blank
then   **pick Primary Units tab and alter:**
           1. Linear Dimensions
              a) Unit Format: Decimal
              b) Precision: 0.00
              c) Decimal separator: '.' Period
              d) Round off: 0

        2. Measurement Scale
          a) Scale factor: 1
        3. Zero Suppression
          a) Trailing: active, i.e. tick
        4. Angular Dimensions
          a) Units Format: Decimal Degrees
          b) Precision: 0.0
          c) Zero Suppression: Trailing active

then   **pick Alternate Units tab and:**
        1. Display alternate units: not active
then   **pick Tolerances tab and:**
        1. Tolerance Format
          a) Method: None
then   **pick OK from New Dimension Style dialogue box**
prompt   Dimension Style Manager dialogue box
with   1. 3DSTD added to styles list
        2. preview of 3DSTD style displayed
        3. description of 3DSTD given
respond 1. pick 3DSTD and it becomes highlighted
        2. pick Set Current
        3. AutoCAD alert perhaps – just pick OK
        4. pick Close

8 Make layer 0 current and menu bar with **Draw-Rectangle** and:
    prompt   Specify first corner point and enter: **0,0 <R>**
    prompt   Specify other corner point and enter: **420,290 <R>**

9 This rectangle will serve as a 'reference base' for our models

10 Menu bar with **View-Zoom-All** and pan to suit

11 Make layer MODEL current

12 Set variables to your own requirements, e.g. GRIPS, PICKFIRST, etc. While I generally work with these off, there will be occasions when they will be toggled on.

13 Menu bar with **File-Save As** and:
    prompt   Save Drawing As dialogue box
    respond 1. scroll and pick named folder (MODR2004)
           2. enter File name: **3DSTDA3**
           3. file type: **AutoCAD 2004 Drawing (*.dwg)**
           4. pick Save

14 Menu bar with **File-Save As** and:
    prompt   Save Drawing As dialogue box
    respond 1. scroll at Files of type
           2. pick **AutoCAD Drawing Template File (*.dwt)**
           3. scroll and pick named folder
           4. enter File name as: **3DSTDA3**
           5. pick Save
    prompt   Template Description dialogue box
    respond 1. Enter: This is my 3D standard sheet
           2. pick OK

15 The created standard sheet has been saved as a drawing file and a template file, both with the name 3DSTDA3. Both have been saved to the MODR2004 named folder – or the name you have given the folder to save all modelling work.

16 *Note*
   a) we could have saved the template file to the AutoCAD Template file – you still can if you want
   b) saving the standard sheet as a template will stop the user 'inadvertently' over-writing the basic 3DSTD standard drawing sheet
   c) all models will be (at present) be created using the 3DSTDA3 template file
   d) all completed models will be saved as drawings to your named folder
   e) the standard sheet has been saved as a drawing file as backup

We are now ready to proceed with creating 3D and solid models.

Chapter 2

2½D models

A 2½D model was one of the first ever 3D displays with a CAD system. It is not a 'real' 3D model and hence the term 2½D was used. While now generally obsolete, the concept is still useful in introducing the new user to the basic 3D terminology.

A 2½D model is created by extruding a 'shape' upwards or downwards from a horizontal plane – called the ELEVATION plane. The actual extruded height (or depth) is called the THICKNESS and can be positive or negative relative to the set elevation plane. This extruded thickness is **always perpendicular** to the elevation plane. The extrusion is in the Z direction of the UCS icon – more on the UCS later. The basic extruded terminology is displayed in Fig. 2.1.

We will demonstrate this type of model with two worked examples.

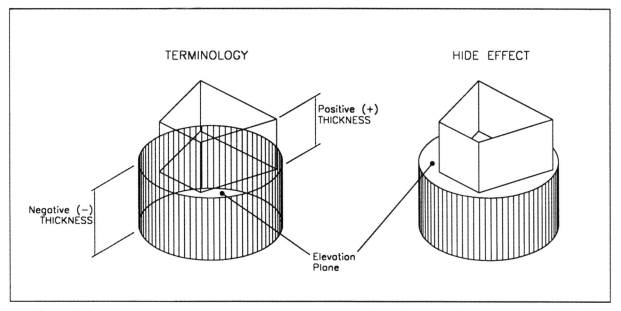

Figure 2.1   Basic extruded terminology.

Example 1

This example is given as a series of user entered steps, these steps also being displayed in Fig. 2.2. The exercise will introduce the user to some of the basic 3D commands and concepts.

To get started:

1   Open your 3DSTDA3 template file and display toolbars to suit, e.g. Draw, Modify and Object Snap

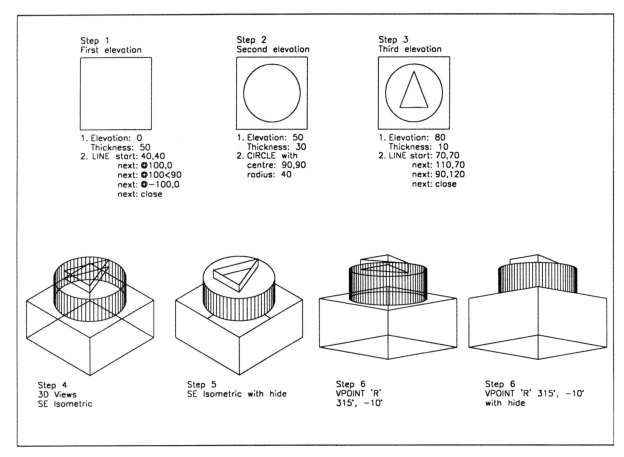

Figure 2.2 Extruded example 1.

2 Layer MODEL should be current

Step 1: the first elevation
1. At the command line enter **ELEV <R>** and:
   prompt Specify new default elevation<0.00> and enter: **0 <R>**
   prompt Specify new default thickness<0.00> and enter: **50 <R>**
2. Nothing appears to have happened?
3. Select the LINE icon and draw:
   Start point: **40,40 <R>**
   Next point: **@100,0 <R>**
   Next point: **@100<90 <R>**
   Next point: **@−100,0 <R>**
   Next point: **C <R>** – the close option
4. A red 'square' will be displayed

Step 2: the second elevation
1. At the command line enter **ELEV <R>** and:
   prompt Specify new default elevation<0.00> and enter: **50 <R>**
   prompt Specify new default thickness<50.00> and enter: **30 <R>**
2. Select the CIRCLE icon and:
   a) centre point: enter **90,90 <R>**
   b) radius: enter **40 <R>**

3. The two objects created are both coloured red as layer MODEL is current. We want to change the colour of the second object (the circle) and at present will use the CHANGE command, so at the command line enter **CHANGE <R>** and:

prompt	Select objects
respond	**pick the circle then right-click**
prompt	Specify change point or [Properties]
enter	**P <R>** – the properties option
prompt	Enter property to change [Color/Elev/Layer/Ltype etc
enter	**C <R>** – the color option
prompt	Enter new color
enter	**green <R>**
prompt	Enter property to change
respond	**right-click and pick Enter**

4. The added circle will now be displayed with a green colour

Step 3: the third elevation
1. With the ELEV command:
   a) set the default elevation to 80
   b) set the default thickness to 10
2. With the LINE icon, draw:
   Start point: **70,70 <R>**
   Next point: **110,70 <R>**
   Next point: **90,120 <R>**
   Next point: **C <R>**
3. With the CHANGE command, change the colour of the three lines to blue, using the same procedure as was used previously
4. We now have a blue triangle inside a green circle inside a red square, and appear to have a traditional 2D plan type drawing
5. Remember that each of the three shapes has been created on a different default elevation plane as follows:
   a) square: elevation 0
   b) circle: elevation 50
   c) triangle: elevation 80

Step 4: viewing the model in 3D
To 'see' the model in 'real' 3D it is necessary to activate the 3D viewpoint command so:
1. From the menu bar select **View-3D Views-SE Isometric**
2. The model will be displayed in 3D. The black 'drawing border' is also displayed in 3D and acts as a 'base' for the model.
3. The orientation of the model is such that it is difficult to know if you are looking down on it, or looking up at it. This is common with 3D modelling and is called **AMBIGUITY**. Another command is required to 'remove' this ambiguity.
4. At this stage save your model with **File-Save As** and ensure:
   a) File type is: AutoCAD 2004 Drawing (*.dwg)
   b) Save in: MODR2004 – your named folder
   c) File name: **EXT-1** – the drawing name
5. This saves the drawing as **C:\MODR2004\EXT-1.dwg** – the path name

Step 5: the hide command
1. From the menu bar select **View-Hide** and the model will be displayed with hidden line removal. It is now easier to visualise.
2. From the screen display it is obvious that the model is being viewed from above, but it is possible to view from different angles.
3. Can you see the difference between the top 'surfaces' of the straight-line objects and the circular object?
4. Menu bar with **View-Regen** to 'restore' the original model.

Step 6: another viewpoint
1. At the command line enter **VPOINT <R>** and:
   prompt   Specify a viewpoint or [Rotate]
   enter   **R <R>** – the rotate option
   prompt   Enter angle in XY plane from X axis and enter: **315 <R>**
   prompt   Enter angle in XY plane and enter: **−10 <R>**
2. The model will be displayed from a different viewpoint without hidden line removal
3. At the command line enter **HIDE <R>**
4. The model will be displayed with hidden line removal and is being viewed from below
5. At the command line enter **REGEN <R>** to restore the original.

Step 7: the shade command
1. Restore the original 3D view with the menu bar sequence **View-3D Views-SE Isometric**
2. Menu bar with **View-Shade-Flat Shaded** and the model will be displayed in colour. This is why the CHANGE command was used earlier in the exercise.
3. Note the icon – more on this later
4. Menu bar with **View-Shade-Gouraud Shaded** and note the effect on the model. Can you observe any difference between the flat shading and the Gouraud shading? Look at the 'cylinder' part of the model.
5. Investigate the other SHADE options available
6. Restore the model to its original display with **View-Shade-2D Wire-frame** and note the icon.

Task

1   With the ERASE command pick any line of the 'base' and a complete 'side' is erased because it is an extrusion

2   Undo the erase effect with **U <R>**

3   Using the erase command pick any point on the top 'circle' and the complete 'cylinder' will be erased

4   Undo this erase effect

5   This completes our first extrusion exercise, so make sure the model is saved, then close the drawing

6   *Note*:
    Although Fig. 2.2 displays several different viewpoints of the model on 'one sheet' this concept will not be discussed until a later chapter. At present you will only display a single viewpoint of the model.

Example 2

This example will use a different method of changing the colour of the model objects – the Properties toolbar so:

1   Open your 3DSTDA3 template file, layer MODEL current and refer to Fig. 2.3

2   At the command line enter **PICKFIRST <R>** and.
    prompt   Enter new value for PICKFIRST
    enter   **1 <R>**
    and   pickfirst box 'attached' to cursor cross-hairs

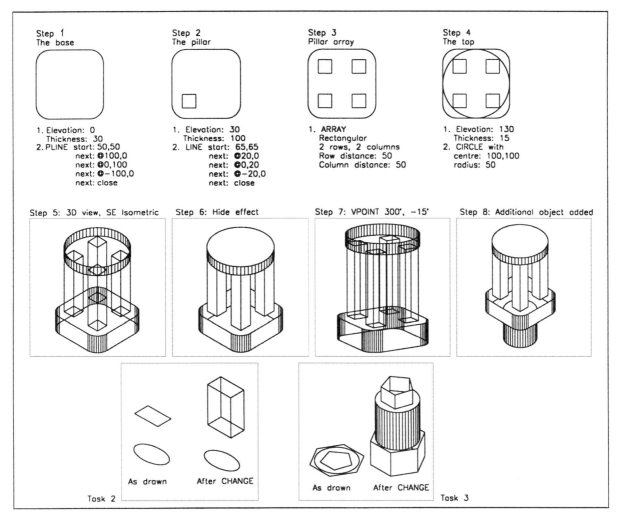

Figure 2.3 Extruded example 2.

Step 1: the base
1. With ELEV at the command line, set the new default elevation to 0 and the new default thickness to 30
2. With the polyline icon from the Draw toolbar, draw a 0 width polyline:
   Start point: **50,50 <R>**
   Next point: **@100,0 <R>**
   Next point: **@0,100 <R>**
   Next point: **@−100,0 <R>**
   Next point: **C <R>**
3. Menu bar with **Modify-Fillet** and:
   prompt   Select first object or [Polyline/Radius/Trim]
   enter   **R <R>** – the radius option
   prompt   Specify fillet radius
   enter   **20 <R>**
   prompt   Select first object [Polyline/Radius/Trim]
   enter   **P <R>** – polyline option
   prompt   Select 2D polyline
   respond **pick any point on the polyline**
4. The red polyline will be filleted at the four corners

Step 2: the first pillar
1. Set the elevation to 30 and the thickness to 100
2. With the LINE command, draw a 20 unit square the lower left corner being at the point 65,65
3. Using the pickbox on the cursor, pick the four lines of the square then pick the Properties icon from the Standard toolbar and:
   prompt   Properties dialogue box
   respond   1. pick the Color line – highlights
                2. pick the scroll arrow at right of Color line
                3. pick Blue – Fig. 2.4
                4. close the Properties dialogue box
                5. press ESC key to deactivate the pickfirst selection
4. The square will be displayed with blue lines

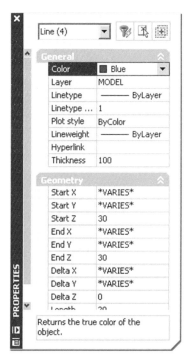

Figure 2.4   The properties dialogue box for the selected four lines.

Step 3: arraying the pillar
1. Select the ARRAY icon from the Modify toolbar and:
   prompt   Array dialogue box
   respond   1. Rectangular Array active
                2. Rows: 2; Columns: 2
                3. Row offset: 50 and Column offset: 50
                4. Angle of Array: 0
                5. pick Select objects and:
   prompt   Select objects at the command line
   respond   window the blue square then right-click
   prompt   Array dialogue box
   respond   pick **Preview<**
   and   blue square arrayed as expected?
   then   Array message and pick **Accept**
2. The blue square will be arrayed in a 2 × 2 matrix pattern

Step 4: the top
1. Set the elevation to 130 and the thickness to 15
2. Draw a circle, centred on 100,100 with radius of 50
3. Using the pickbox:
   a) pick the circle then the Properties icon
   b) set the colour to green

Step 5: the 3D viewpoint
1. Menu bar with **View-3D Views-SE Isometric**
2. The model is displayed in 3D but appears rather 'cluttered'

Step 6: hiding the model
1. Menu bar with **View-Hide** model displayed with hidden line removal
2. Menu bar with **View-Regen** to restore the original model

Step 7: setting another viewpoint
1. At the command line enter **VPOINT <R>** and:
   prompt Specify a new viewpoint or [Rotate]
   enter **R <R>** – the rotate option
   prompt Enter angle in XY plane from X axis and enter: **300 <R>**
   prompt Enter angle from XY plane and enter: **−15 <R>**
2. Menu bar with **View-Hide** to 'see' the model from below
3. Menu bar with **View-Regen** to restore the original model
4. Restore the original 3D view with **View-3D Views-SE Isometric**

Step 8
1. The model should be displayed in 3D at a SE Isometric viewpoint
2. Using the command line, set the elevation to 0 and the thickness to −60
3. Draw a circle with centre at 100,100 and radius 30
4. The circle will be displayed in 3D as a 'cylinder'
5. Change the colour of the added 'cylinder' to magenta
6. As the model is complete, save as **C:\MODR2004\EXT-2**.

Task 1

Use the menu bar with the following menu bar sequences:
a) View-3D Views-SE Isometric
b) View-Hide and note green circle display
c) View-Shade-Flat Shaded and note colour effect and icon
d) View-Shade-3D Wire-frame
e) View-Hide and note the green circle display
f) View-Shade-Flat Shaded, Edges On
g) View-Shade-2D Wire-frame and note the green circle display
h) View-Regen to 'restore' the original model.

Task 2

1   Still with the SE Isometric viewpoint displayed

2   Set the elevation to 0 and the thickness to 100

3   With **Draw-Rectangle** create a rectangle anywhere on the screen

4   With **Draw-Ellipse-Center** create an ellipse anywhere on the screen

5   Both the rectangle and the ellipse will be displayed without any thickness, although the thickness was set to 100 in step 2

6 At the command line enter **CHANGE <R>** and:
   prompt   Select objects
   respond pick any point on the rectangle then right-click
   prompt   Specify change point or [Properties]
   enter   **P <R>** – the Properties option
   prompt   Enter property to change [Color/Elev/Layer etc
   enter   **T <R>** – the thickness option
   prompt   Specify new thickness <0.00>
   enter   **100 <R>**
   prompt   Enter property to change
   enter   **<R>** – to end command as no other properties to change

7 The rectangle will now be displayed in 3D with a thickness

8 Using the same sequence and entries as step 6, select the ellipse. No thickness will 'be added'

9 With the CHANGE command, alter the elevation of the ellipse to 50 – any change to the ellipse. Now try using the Properties toolbar to change the thickness of the ellipse.

Task 3

1 Display a SE Isometric viewpoint and set the elevation and thickness both to 0. Layer MODEL still current

2 Draw the following objects (panning to suit):
   a) polygon with 6 sides, centred on 0,0 and inscribed in a 50 radius circle
   b) circle, centre on 0,0 with radius 40
   c) polygon with 5 sides, centred on 0,0 and inscribed in a 30 radius circle

3 Using the Properties dialogue box (with PICKFIRST set to 1), alter the properties of the three objects using the following information:

object	elev	thickness	colour
6 sided polygon	0	50	red
circle	50	80	blue
5 sided polygon	130	30	green

4 Investigate the hide and shade commands and other 3D viewpoints

5 This exercise is now complete. Do not save these additions.

Summary

1 An extruded model is created from an elevation and thickness

2 Extruded models are created 'as sides'

3 The elevation and thickness values are usually set from the command line

4 The elevation and thickness of objects can be altered with:
   a) command line CHANGE with PICKFIRST 0
   b) Properties icon with PICKFIRST 1 – dialogue box method

5 Extruded models are viewed in 3D with the 3D Views command which will be discussed in detail in a later chapter

6 2½D models are displayed with AMBIGUITY, i.e. are you looking down from the top or up from the bottom?

7 The HIDE command is used to display 2½D models with hidden line removal. This removes the AMBIGUITY effect

8 The SHADE command gives useful displays with coloured objects.

Assignment

At the end of several chapters, the reader will be asked to complete certain assignments/activities. These will reinforce the skills you have learned during the chapter. During these assignments you will frequently meet a character called MACFARAMUS. This august gentleman was considered to be a great architect, engineer, scientist in times gone by, but sadly most of his works have not been given the credit they deserve.

Your first assignment is to create as a 2½D model, a famous artefact of MACFARAMUS which consists basically of trimmed circles to form a CAM type outline. This artefact was discovered on a sheet of papyrus, the sizes having absolutely no relation to modern day metric sizes. The explorers who found the 'drawing' stated that the sizes appeared to be in CRATURS, which were meaningless to them (and to me). These sizes have been altered to metric, to make the drawing easier to complete.

Note that all the activity drawings are at the end of the book, starting on page 327.

Activity 1: CAM artefact of MACFARAMUS

1 Open your 3DSTDA3 template file

2 Using the sizes given, create the basic outline from trimmed circles, or any other method of your choice

3 When the basic outline has been completed, convert it into a single polyline

4 Save this polyline outline to your named folder as **CAM** as it will be used for other activities

5 Now create the 2½D model, using the layout sizes given

6 Decide for yourself whether to:
   a) set the elevation and thickness values then draw the shapes
   b) draw the shapes with elevation 0 then change the elevation and thickness values

7 Decide on whether to use the CHANGE or Properties dialogue box, i.e.
   a) PICKFIRST 0 – CHANGE at command line
   b) PICKFIRST 1 – Properties dialogue box

8 When the model is complete, view at different 3D viewpoints then hide and shade

9 Note that at present you will not be able to obtain the two different views on the one screen (unless you have some prior AutoCAD 3D knowledge)

10 Remember to save the completed model.

Chapter 3

The UCS and 3D co-ordinates

The AutoCAD draughting package allows the user two co-ordinates systems, these being:

1   the world co-ordinate system (**WCS**)
2   the user co-ordinate system (**UCS**)

The WCS

All readers should be familiar with the basic 2D co-ordinate concept of a point described as P1 (30,40) – Fig. 3.1. Such a point has 30 units in the positive X direction and 40 units in the positive Y direction. These ordinates are relative to an XY axes system with the origin at the point (0,0). This origin is normally positioned at the lower left corner of the screen and is perfectly satisfactory for 2D draughting but not for 3D modelling.

Drawing in 3D requires a third axis (the Z axis) to enable 3D co-ordinates to be used. The screen monitor is a flat surface and it is difficult to display a three-axis co-ordinate system on it. AutoCAD overcomes this difficulty by using an **ICON** and this icon can be moved to different positions on the screen and can be orientated on existing objects.

Figure 3.2 shows the basic idea of how the icon has been constructed. The X and Y axes are displayed in their correct orientations while the Z axis is pointing outwards towards the user. The **W** on the icon indicates that the user is working with the WCS. The origin is at the point (0,0,0) and is positioned at the lower left corner of the screen – as it is in 2D. The status bar displays the three co-ordinates of any point on the screen, but these figures can be misleading, especially when viewing in 3D. The origin point can be positioned to suit the model being created – more on this later.

The point P2 (30,40,5) is thus defined as 30 units in the positive X direction, 40 units in the positive Y direction and 50 units in the positive Z direction. Similarly the point P3 (−40,−50,−30) has 40 units in the negative X direction, 50 units in the negative Y direction and 30 units in the negative Z direction.

In the previous chapter, all the extruded models were created with the WCS.

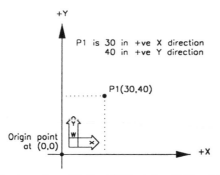

Figure 3.1   2D co-ordinate entry with the WCS at the (0,0) origin.

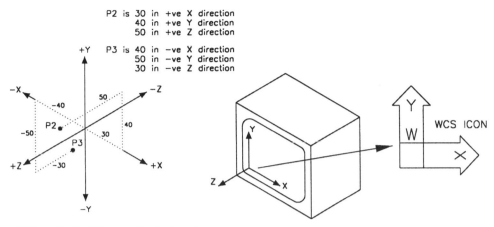

Figure 3.2   3D co-ordinate input.

The UCS

The UCS is one of the most important concepts in 3D modelling and all users must be fully conversant with it. The user co-ordinate system allows the operator several options including:

a) setting a new UCS-origin point
b) moving the origin to any point (or object) on the screen
c) aligning the UCS icon with existing objects
d) aligning the UCS icon to suit any 'plane' on a model
e) rotating the icon about the X, Y and Z axes
f) saving UCS 'positions'
g) recalling previously saved UCS settings

Icon display

AutoCAD 2004 allows the user to display the icon as a 2D symbol or as a 3D symbol. The previous discussion has assumed that the user has the traditional AutoCAD 2D icon displayed (as Fig. 3.2) but this may not be the icon displayed on your screen. To investigate the UCS icon display:

1   Close all existing drawings and refer to Fig. 3.3

2   Start a new metric drawing from scratch to display a blank screen

3   Menu bar with **View-Display-UCS Icon** and from the cascade effect:
    a) ensure On active – tick
    b) ensure Origin active – tick
    c) pick Properties and:
      prompt   UCS Icon dialogue box
      respond 1. UCS icon style: pick 2D and note Preview
               2. UCS icon size: set to suit, e.g. 12
               3. UCS icon color: set to suit (Black default)
               4. Layout tab icon color: set to suit (Black default)
               5. pick OK

4   The icon is displayed as Fig. 3.3(a)

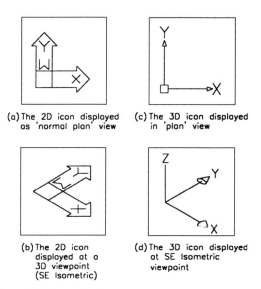

(a) The 2D icon displayed as 'normal plan' view   (c) The 3D icon displayed in 'plan' view

(b) The 2D icon displayed at a 3D viewpoint (SE Isometric)   (d) The 3D icon displayed at SE Isometric viewpoint

Figure 3.3   The 2D and 3D icon display.

5   Menu bar with **View-3D Views-SE Isometric** and the icon will be displayed in 3D as Fig. 3.3(b)

6   Enter **U <R>** to restore the original 'plan' icon

7   Repeat step 3 and from the UCS Icon dialogue box:
    a) set UCS icon style: pick 3D and note Preview
    b) ensure Cone active – tick
    c) set Line width: 1
    d) dialogue box as Fig. 3.4
    e) pick OK

8   The icon will be displayed as Fig. 3.3(c) and as Fig. 3.3(d) if a SE Isometric viewpoint is set

As the user, you must now decide on whether to display the 2D or 3D icon. It is your preference.

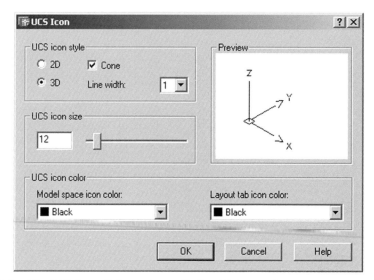

Figure 3.4   The UCS Icon dialogue box with the 3D icon set.

UCS icon exercise

The appearance of the co-ordinate icon alters depending on:
a) its orientation, i.e. how it is 'attached' to objects
b) the viewpoint selected or entered

To investigate the UCS icon display, the following exercise is given as a sequence of operations which the reader should complete. No drawing is involved and it should be noted that several of the commands will be new to some readers, all of which will be explained later. The object of the exercise is to make the reader aware of the 'versatility' of the co-ordinate icon.

1   Close all existing drawings then open your 3DSTDA3 template file and refer to Fig. 3.5

2   Menu bar with **View-Display-UCS icon** and:
a) On and Origin both active, i.e. tick
b) pick Properties and activate the 2D UCS icon style

3   The icon displayed at the lower left corner of the screen has a W on it, indicating that it is the WCS icon as Fig. 3.5(a). This is the 'normal' default icon.

4   Select the PAN icon from the Standard toolbar or enter PAN <R> at the command line and:
a) pan the screen upwards and to the right
b) right-click and pick Exit

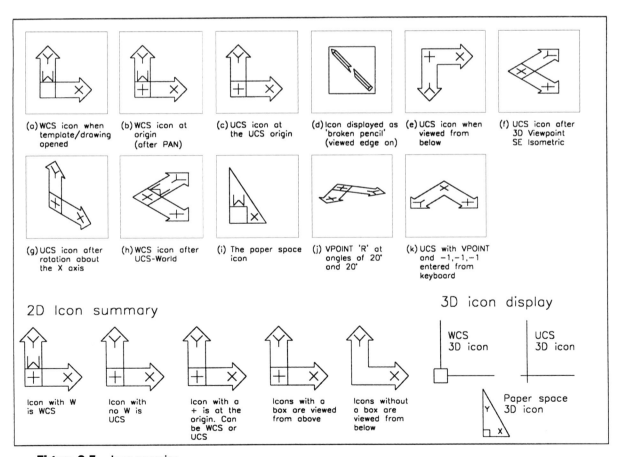

Figure 3.5   Icon exercise.

5   The icon will be displayed as Fig. 3.5(b) and be positioned at the lower left corner of the 'drawing sheet'. It has a + sign added in the 'box', indicating that the icon is positioned at the origin.

6   With snap on, move the cursor onto the icon + and observe the status bar – the co-ordinates should be 0.00, 0.00, 0.00

7   Pick the Undo icon from the Standard toolbar to restore the icon to its original position – Fig. 3.5(a)

8   Menu bar with **Tools-New UCS-Origin** and:
    prompt   Specify new origin point<0,0,0>
    enter   **100,100 <R>**
    and   the icon moves to the entered point and is displayed as Fig. 3.5(c). It has no W indicating that it is a UCS icon and has a + indicating it is at the origin

9   With snap on, move the cursor onto the + and observe the status bar co-ordinates in the status bar. They should all display 0.00.

10   Menu bar with **Tools-New UCS-X** and:
     prompt   Specify rotation angle about X axis<0.0>
     enter   **90 <R>**
     and   icon displayed as Fig. 3.5(d). This is the AutoCAD 'broken pencil' icon indicating that we are looking at it 'edge-on'

11   At the command line enter **UCS <R>** and:
     prompt   Enter an option [New/Move/..
     enter   **N <R>** – the new option
     prompt   Specify origin of new UCS or [ZAxis/3point/..
     enter   **X <R>** – the rotate about X axis option
     prompt   Specify rotation angle about X axis
     enter   **90 <R>**
     and   icon displayed as Fig. 3.5(e) and is being viewed from below – there is no 'box'. The + is still displayed indicating the UCS icon is still at the origin.

12   Menu bar with **Tools-New UCS-X** and enter 180 as the rotation angle. The icon will again be displayed as Fig. 3.5(c)

13   Menu bar with **View-3D Views-SE Isometric** and the icon will be displayed in 3D as Fig. 3.5(f). It still has a + and is therefore still at the origin.

14   At the command line enter **UCS <R>** and:
     prompt   Enter an option
     enter   **N <R>** then **X <R>** – new and X rotate options
     prompt   Specify rotation angle about X axis
     enter   **90 <R>**
     and   icon displayed as Fig. 3.5(g)

15   Undo the UCS-X rotation with U <R> or pick the Undo icon to display the icon as Fig. 3.5(f) again

16   At the command line enter **ZOOM <R>** then **0.75 <R>** to 'reduce' the scale of the drawing sheet

17   Menu bar with **Tools-New UCS-World** and the icon will be displayed as Fig. 3.5(h). This is a WCS icon positioned at the 'world' origin point – the lower left corner of the 'drawing sheet'. The icon is still displayed in 3D.

18   Menu bar with **View-3D Views-Plan View-World UCS** and the icon should be as the original Fig. 3.5(a). The screen should display the drawing sheet 'as opened'.

19 Left click on the word MODEL in the status bar and:
    prompt   `Page Setup dialogue box`
    respond **pick Cancel** – more on this later
    and   the icon will be displayed as Fig. 3.5(i). This is the paper space icon which
             will be discussed in more detail in a later chapter.

20 At present undo this paper space effect with U <R> to restore the icon as Fig. 3.5(a)

21 Enter/select the following sequences:
    a) Menu bar with Tools-New UCS-Origin and enter 100,100 to display the icon as
        Fig. 3.5(c)
    b) enter VPOINT <R> then R <R> with angles of 20 and 20 and the icon will be dis-
        played as Fig. 3.5(j)
    c) enter VPOINT <R> then −1,−1,−1 to give the icon as Fig. 3.5(k)
    d) enter VPOINT <R> then R <R> with angles of 0 and 90 – Fig. 3.5(c)
    e) enter UCS <R> then W <R> – Fig. 3.5(a)

22 This completes the first part of the icon exercise

23 *Note*: we could have used the UCS toolbar with icons during this exercise, but at this
    stage I think that the menu bar and command line selections give the user a 'better
    understanding' of what is actually happening. You can investigate the UCS toolbar for
    yourself.

24 *Icon summary*
    Figure 3.5 displays a summary of the various 2D icons which can be displayed on the
    screen. These are:
    a) icon with a W is a WCS icon
    b) icon with no W is a UCS icon
    c) icon with a + is at the origin
    d) icon with a 'box' is viewed from above
    e) icon with no 'box' is viewed from below

25 *Task*
    a) with the UCS Icon dialogue box, set a 3D style icon
    b) repeat the steps in the previous exercise and observe the orientation of the 3D icon
    c) generally the same 'type of orientation' is obtained with the 3D icon as with the 2D
        icon. The paper space icon with the 3D style is slightly different from the 2D icon.
    d) the WCS and UCS icons with a 3D style setting are displayed in Fig. 3.5
    e) it is user-preference whether to use a 2D or 3D icon

26 This exercise is now complete.

Orientation of the UCS

The completed exercise has demonstrated that the UCS icon can be moved to any
point on the screen and rotated about the three axes (we only used the X axis rotation, but the procedure is the same for the Y and Z axes). It is thus important for the
user to be able to determine the correct orientation of the icon, i.e. how the X, Y and
Z axes are configured in relation to each other.

The axes orientation is determined by the **right-hand rule** and is demonstrated in
Fig. 3.6. The knuckle of the right hand is at the origin and the position of the thumb,
index finger and second finger determine the direction of the positive X, Y and Z axes
respectively.

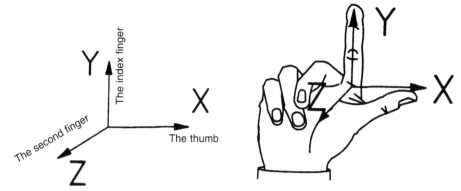

Figure 3.6   The right-hand rule.

3D co-ordinate input

Co-ordinate input is generally required at some time during the creation of a 3D model. With 3D draughting there are three types of co-ordinates available, each having both absolute and relative entry modes. The three co-ordinate types with their formats and examples are:

Type	*Format*	*Absolute*	*Relative*
Cartesian	x dist,y dist,z dist	100,150,120	@300,−100,−50
Cylindrical	dist<angle,Z dist	150<55,120	@75<−15,−120
Spherical	dist<angle 1<angle 2	80<30<50	@120<−10<75

To investigate the different types of co-ordinate input we will draw some objects on the screen. We will also investigate the effect of the icon position on the co-ordinate entries.

1   Close any existing drawings

2   Open your 3DSTDA3 template file and refer to Fig. 3.7

3   Menu bar with **View-Display-UCS Icon** and:
   a) On and Origin – tick
   b) Properties and set a 2D style
   c) These selections ensure that the icon is displayed in 2D on the screen and is always 'positioned' at the origin point

4   Menu bar with **View-3D Views-SE Isometric** to display the screen in 3D

5   Menu bar with **View-Zoom-Scale** and:
   prompt   Enter a scale factor and enter: **0.75 <R>**

6   The WCS icon should be positioned at the left vertex of the black border – point A in Fig. 3.7

7   Make three new layers – L1, L2, L3 with continuous linetype and colour numbers 30, 72, 240 respectively

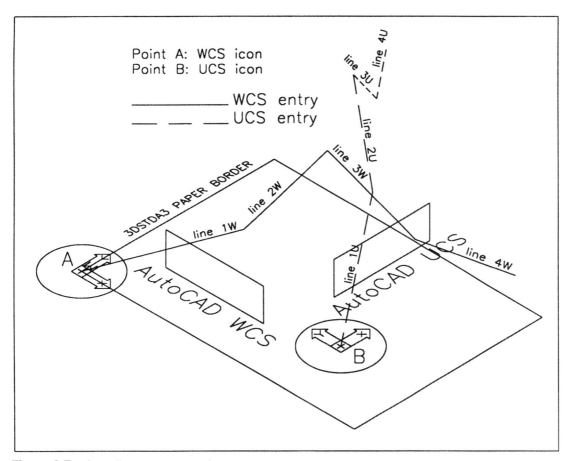

Figure 3.7 Co-ordinate entry exercise.

A) WCS entry

1 With layer L1 current, use the LINE icon and draw:

First point	**0,0,0 <R>**		
Next point	**150,100,80 <R>**	absolute	line 1W
Next point	**@50,80,90 <R>**	relative absolute	line 2W
Next point	**@100<30,−100 <R>**	relative cylindrical	line 3W
Next point	**@120<40<−20 <R>**	relative spherical	line 4W
Next point	**right-click and pick Enter**		

2 Draw a circle, centre: 0,0,0 with radius: 50

3 Add the following item of text:
   a) start point: 40,40,0
   b) height: 20 with 0 rotation
   c) item: AutoCAD WCS

4 a) With ELEV at the command line, set the current elevation to 0 and the current thickness to 50
   b) Draw a line from 60,70 to @150,0,0

5 Set the elevation and thickness values back to 0

B) UCS entry

1 Menu bar with **Tools-New UCS-Origin** and:
   prompt Specify new origin point
   enter **300,100,0 <R>**

2 Menu bar with **Tools-New UCS-Z** and:
   prompt Specify rotation angle about Z axis
   enter **90 <R>**

3 The icon should be positioned and orientated at point B

4 With layer L2 current, use the LINE icon and draw:
   First point **0,0,0 <R>**
   Next point **150,100,80 <R>**   absolute   line 1U
   Next point **@50,80,90 <R>**   relative absolute   line 2U
   Next point **@100<30,−100 <R>**   relative cylindrical line 3U
   Next point **@120<40<−20 <R>**   relative spherical   line 4U
   Next point **right-click and Enter**

5 Menu bar with **View-Zoom-All** to 'see' the additional lines

6 Draw a circle, centred on 0,0,0 with a 50 radius

7 Add the text item:
   a) start point: 40,40,0
   b) height: 20 with 0 rotation
   c) item: AutoCAD UCS

8 *a*) With ELEV at the command line, set the current elevation to 0 and the current thickness to 50
   b) Draw a line from 60,70 to @150,0

9 Set the elevation and thickness values back to 0

C) WCS entry with UCS icon

1 The UCS should still be at position B

2 Ensure elevation and thickness set to 0 and make layer L3 current

3 With the LINE icon draw:
   First point ***0,0,0 <R>**
   Next point ***150,100,80 <R>**
   Next point **@*50,80,90 <R>**
   Next point **@*100<30,−100 <R>**
   Next point **@*120<40<−20 <R>**
   Next point **right-click and Enter**

4 These lines should be identical to those created on layer L1 when the WCS was current.

Task

1 Save the co-ordinate exercise if required, but we will not refer to it again

2 With **File-Open** recall your 3DSTDA3 template file

3 Menu bar with **View-Display-UCS icon** and ensure:
   a) On and Origin both active – tick
   b) Properties and set icon to your preference 2D or 3D

4   Menu bar with **View-3D Views-SE Isometric**

5   Menu bar with **View-Zoom-Scale** and enter a factor of **0.75**

6   The WCS icon should be positioned at left vertex of the border

7   Save this layout to your named folder as:
    a) the **3DSTDA3.dwt** template file and replace the existing template file; enter a suitable template description
    b) the **3DSTDA3.dwg** drawing file, overwriting the existing file

8   This will allow the template file to opened in 3D with the icon always 'set' to the origin position.

Summary

1   There are two co-ordinate systems:
    a) the WCS
    b) the UCS

2   Each system has its own icon

3   The WCS is a fixed system, the origin being at 0,0,0

4   The WCS icon is 'standard' and does not alter in appearance. The WCS icon is denoted with the letter W

5   The UCS system allows the user to define the origin, either as a point on the screen or referenced to an existing object

6   The UCS icon alters in appearance dependent on the viewpoint

7   The UCS icon can be rotated about the three axes

8   The UCS current position can be saved and recalled

9   The user can set a 2D or 3D UCS icon style

10 3D co-ordinate input can be:
    a) Cartesian, e.g. 10,20,30
    b) Cylindrical, e.g. 10<20,30
    c) Spherical, e.g. 10<20<30

11 Both absolute and relative modes of input are possible with the three 'types' of co-ordinates, e.g.
    a) absolute cylindrical: 100<200,50
    b) relative cylindrical: @100<200,50

12 3D co-ordinate input can be relative to the current UCS position or to the WCS, e.g.
    a) 100,200,150 for UCS entry
    b) *100,200,150 for WCS entry

13 It is recommended that **3D co-ordinate input is relative to the current UCS position**.

Chapter 4

Creating a 3D wire-frame model

In this chapter we will create a 3D wire-frame model and use it to:
a) investigate how the UCS can be set and saved
b) add 'objects' and text to the model 'planes'
c) modify the model.

Getting started

1 Open your 3DSTDA3 template file to display:
 a) a SE Isometric viewpoint with the black border
 b) the WCS icon at the left vertex of the border

2 Ensure layer MODEL is current and refer to Fig. 4.1

3 Display the Draw, Modify and Objects Snap toolbars.

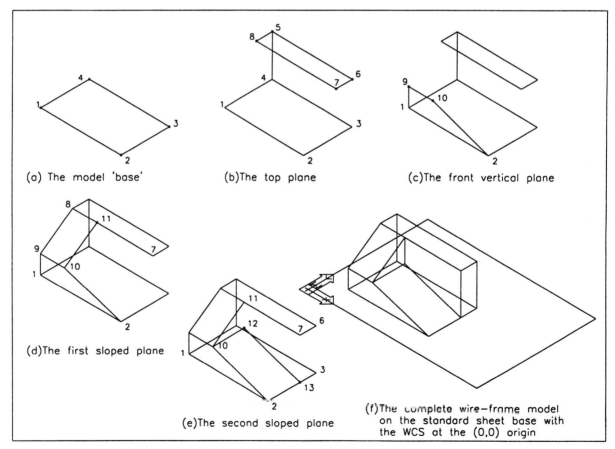

Figure 4.1   Construction of the wire-frame model 3DWFM.

Creating the wire-frame model

1 To create the base of the model – Fig. 4.1(a), select the LINE icon and draw:

Start point	**50,50 <R>**	pt1
Next point	**@200,0 <R>**	pt2
Next point	**@0,120 <R>**	pt3
Next point	**@−200,0 <R>**	pt4
Next point	**close**	

2 The top plane – Fig. 4.1(b) is also created from lines, so with the LINE icon draw:

Start point	**Intersection icon of pt4**	
Next point	**@0,0,100 <R>**	pt5
Next point	**@200,0,0 <R>**	pt6
Next point	**@0,−40,0 <R>**	pt7
Next point	**@−200,0,0 <R>**	pt8
Next point	**Intersection icon of pt5**	pt5
Next point	right-click and pick Enter	

3 If you cannot 'see' the complete model, then menu bar with **View-Zoom-Scale** and enter a scale factor to suit, e.g. 0.9

4 To create the front vertical plane – Fig. 4.1(c), select the LINE icon and draw:

Start point	**Intersection icon of pt1**	
Next point	**@0,0,45 <R>**	pt9
Next point	**@60,0,0 <R>**	pt10
Next point	**Intersection icon of pt2**	
Next point	right-click and Enter	

5 With the LINE icon draw:

Start point	**Intersection of pt9**
Next point	**Intersection of pt8** then right-click/Enter

6 LINE icon again:

Start point	**Intersection of pt10**	
Next point	**Perpendicular to line 78**	pt11
Next point	right-click and Enter	
and	first sloped plane created – Fig. 4.1(d)	

7 To create the second sloped plane – Fig. 4.1(e), select the LINE icon and draw:

Start point	**Intersection of pt10**	
Next point	**@0,80,0 <R>**	pt12
Next point	**Perpendicular to line 23**	pt13
Next point	right-click and Enter – Fig. 4.1(e)	

8 To completing the model, three lines require to be added, so with the LINE icon draw:

  a) from pt3 to pt6

  b) from pt7 to pt13

  c) from pt11 to pt12

9 The completed model is displayed in Fig. 4.1(f) on 'its base', i.e. the standard sheet black border.

10 At this stage save the model as a drawing file with the name **C:\MODR2004\3DWFM** or your named folder

11 *Note*:
The model has been created using 3D co-ordinate input with the WCS, i.e. no attempt has been made to use the UCS. This is a perfectly valid method of creating wire-frame models, but difficulty can be experienced if objects and text have to be added to the various 'surfaces' of the model when the co-ordinates need to be calculated. Using the UCS usually overcomes this type of problem.

Moving around with the UCS

To obtain a better understanding of the UCS and how it is used with 3D models, we will use the created wire-frame model to add some objects and text. The sequence is quite long but it is important that you persevere and complete the exercise. Both menu bar and keyboard entry methods will be used to activate the UCS command.

1   Open the wire-frame model **C:\MODR2004\3DWFM** or continue from the previous exercise. This model has the WCS icon at the black border origin point – the left vertex

2   Menu bar with **View-Display-UCS Icon** and:
    a) On and Origin both active (tick)
    b) select Properties and set a 2D UCS icon style

3   Refer to Fig. 4.2

4   PAN the layout until the lower black border vertex is near the lower edge of the screen. This will allow us to 'see' any UCS movements more clearly.

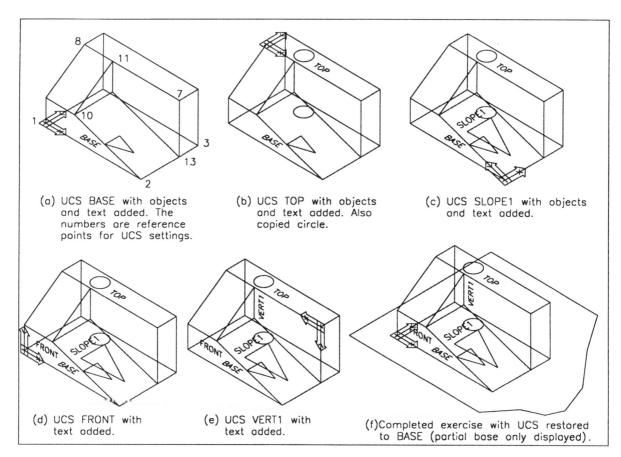

(a) UCS BASE with objects and text added. The numbers are reference points for UCS settings.

(b) UCS TOP with objects and text added. Also copied circle.

(c) UCS SLOPE1 with objects and text added.

(d) UCS FRONT with text added.

(e) UCS VERT1 with text added.

(f) Completed exercise with UCS restored to BASE (partial base only displayed).

Figure 4.2   Investigating the UCS and adding objects and text to 3DWFM.

5 Menu bar with **Tools-New UCS-Origin** and:
prompt   Specify new origin point <0,0,0>
respond **Intersection icon and pick pt1**
and   *a*) icon 'moves' to selected point – Fig. 4.2(a)
          b) it is a UCS icon: there is no W
          c) it is at the origin: there is a +
note   if the icon does not move to the selected point, menu bar with **View-Display-UCS Icon** and pick activate Origin

6 Now that the icon has been repositioned at point 1, we want to save its 'position' for future recall, so at the command line enter **UCS <R>** and:
prompt Enter an option
enter   **S <R>** – the save option
prompt Enter name to save current UCS
enter   **BASE <R>**

7 Make layer OBJECTS current and use the LINE icon to draw:
Start point **100,25,0 <R>**
Next point **@0,30,0 <R>**
Next point **145,40,0 <R>**
Next point **close**

8 Make layer TEXT current and menu bar with **Draw-Text-Single Line Text** and:
a) start point: 60,10,0
b) height: 10 and 0 rotation
c) text item: BASE

9 The line objects and text item are added as Fig. 4.2(a)

10 Menu bar with **Tools-New UCS-Origin** and:
prompt   Specify new origin point <0,0,0>
respond **Intersection icon and pick pt8**
and   icon 'jumps' to the selected point – Fig. 4.2(b)

11 At the command line enter **UCS <R>** and:
prompt Enter an option
enter   **S <R>** – the save option
prompt Enter name to save current UCS
enter   **TOP <R>**

12 With layer OBJECTS current draw a circle with centre: 60,20 and radius: 15

13 With layer TEXT current, add single line text using:
a) start point: 85,10
b) height: 10 with 0 rotation
c) text item: TOP

14 Using the COPY icon:
a) select objects: pick the circle then right-click
b) base point: Center icon and pick the circle
c) second point: enter @0,0,−100 <R> – Fig. 4.2(b)
d) question: why these co-ordinates?

15 Menu bar with **Tools-UCS-3Point** and:
prompt   Specify new origin point<0,0,0>
respond **Endpoint icon and pick pt2**
prompt   Specify point on positive portion of the X-axis
respond **Endpoint icon and pick pt3**
prompt   Specify point on positive-Y portion of the UCS XY plane
respond **Endpoint icon and pick pt10**

16 The UCS icon will move to point 2 and be 'aligned' on the sloped surface as Fig. 4.2(c)

17 *Note*:
    The 3 point option of the UCS command is 'asking the user' for three points to define the UCS icon orientation, these being:
    1. *first prompt*   the origin point
    2. *second prompt*   the direction of the X axis
    3. *third prompt*   the direction of the Y axis

18 Save this UCS position by entering at the command line **UCS <R>** then **S <R>** and:
    prompt Enter name to save current UCS
    enter   **SLOPE1 <R>**

19 With layer OBJECTS current use the LINE icon to draw:
    Start   **15,100,0**
    Next   **@50,0,0**
    Next   **40,30,0**
    Next   **close**

20 With layer TEXT current, add a single text item using:
    a) start point: centred on 10,110
    b) height: 10 with 0 rotation
    c) item: SLOPE1 – Fig. 4.2(c)

21 At command line enter **UCS <R>** and:
    prompt Enter an option
    enter   **R <R>** – the restore option
    prompt Enter name of UCS to restore
    enter   **BASE <R>**
    and   icon restored to the base point as Fig. 4.2(a)
    (The restore option is used extensively with UCS's)

22 Menu bar with **Tools-New UCS-X** and:
    prompt Specify rotation angle about X axis
    enter   **90 <R>**
    and   icon displayed as Fig. 4.2(d)

23 At command line enter **UCS <R>** then **S <R>** for the save option and **FRONT <R>** as the UCS name to save

24 With layer TEXT current add an item of text with:
    a) start point: 25,20
    b) height: 10 with 0 rotation
    c) text: FRONT – Fig. 4.2(d)

25 Menu bar with **Tools-New UCS-3 Point** and:
    prompt Specify new origin point
    respond **Intersection icon and pick pt7**
    prompt Specify point on positive portion of the X-axis
    respond **Intersection and pick pt11**
    prompt Specify point on positive-Y portion of the UCS XY plane
    respond **Intersection icon and pick pt13**

26 The UCS icon will be aligned as Fig. 4.2(e)

27 Save this UCS position as VERT1 – easy? (UCS-S-VERT1)

28 With layer TEXT current add a text item with:
    a) start point: 120,50
    b) height: 10
    c) rotation: −90
    d) text: VERT1 – Fig. 4.2(e)

29 Restore UCS BASE and the model will be displayed as Fig. 4.2(f)

30 Make layer MODEL current and save the drawing at this stage as **C:\MODR2004\ 3DWFM** updating the original wire-frame model

31 *Note*:
    The various UCS positions have been saved and recalled by entering **UCS** at the command line. There is a dialogue box method of saving and recalling UCS positions. This will be discussed in the next chapter.

Modifying the wire-frame model

To further investigate the UCS we will modify the wire-frame model, so refer to Fig. 4.3 and:

1 3DWFM still on the screen? – if not open the drawing file

2 Layer MODEL current with UCS BASE – Fig 4.3(a)

3 Select the CHAMFER icon from the Modify toolbar and:
   a) set both chamfer distances to 30
   b) chamfer lines 7–11 and 7–13
   c) chamfer lines 5–6 and 6–3

4 Now add two lines to complete the 'chamfered corner' and erase the unwanted original corner line – Fig. 4.3(b).

5 Restore UCS VERT1 and note its position – Fig. 4.3(c)

6 Draw two circles:
   a) centre at 80,0,0 with radius 30
   b) centre at 80,0,−40 with radius 30 – Fig. 4.3(c)

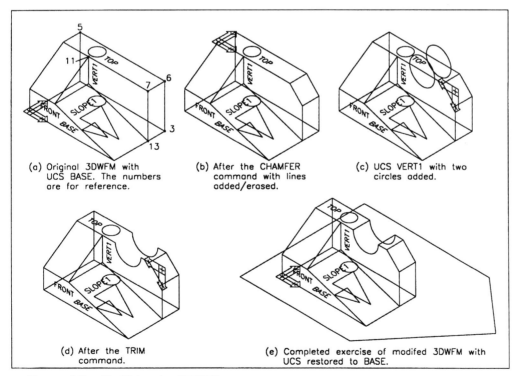

(a) Original 3DWFM with UCS BASE. The numbers are for reference.

(b) After the CHAMFER command with lines added/erased.

(c) UCS VERT1 with two circles added.

(d) After the TRIM command.

(e) Completed exercise of modifed 3DWFM with UCS restored to BASE.

Figure 4.3 Modifying the 3DWFM.

7 Using the TRIM icon from the Modify toolbar:
   a) trim the two circles 'above' the model
   b) trim the two lines 'between' the circles – Fig. 4.3(d)

8 Move the TOP text item from: ENDPOINT of pt5, by: @80,0

9 Draw in the two lines on the top plane and restore UCS BASE

10 The modified model is now complete – Fig. 4.3(e)

11 Save the model as **C:\MODR2004\3DWFM** updating the existing model drawing

12 *Note*:
    The user should realise that the UCS is an important concept with 3D modelling. Indeed
    I would suggest that 3D modelling would be very difficult (if not impossible) without it.

Task 1

1 The wire-frame model has eleven flat planes and one 'curved surface'. We have set and
   saved UCS positions for five of these planes – BASE, TOP, SLOPE1, FRONT and VERT1.

2 You now have to set and save the other six flat UCS positions, i.e. one for each surface
   and add an appropriate text item to that surface.

3 My suggestions for the UCS name and text item are LEFT, RIGHT, REAR, SLOPE2,
   SLOPE3 and VERT2 but you can use any names that you consider suitable.

4 Fig. 4.4 displays the complete wire-frame model with text added to every plane (with
   the exception of the curved surface) using the UCS positions I 'set'. Realise that your
   additional text may differ in appearance from mine. This is acceptable as your UCS
   positions may be 'set' different from mine.

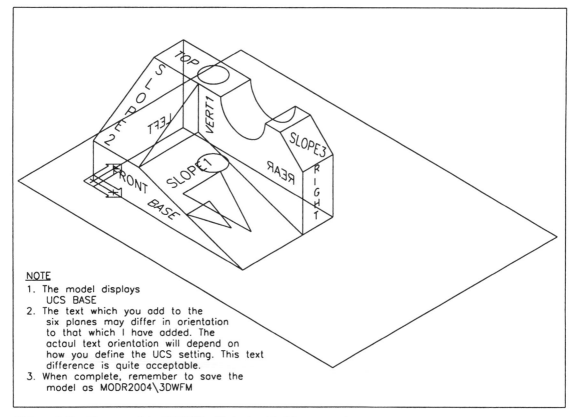

Figure 4.4   The complete 3DWFM with text added to every plane.

5   When complete, remember to save as **MODR2004\3DWFM** as it will be used in other chapters.

Task 2

1   Restore UCS BASE – should be current?

2   With the MOVE command:
    a) window the complete model then right-click
    b) base point: 0,0
    c) second point: @100,100

3   The complete model moves as expected, but do the set UCS's move with the model? This can be a nuisance when moving models. The UCS is 'not tied' to a specific model, it is **ONLY A POSITION ON THE SCREEN**

4   This exercise is now complete. Do not save the changes.

Summary

1   Wire-frame models are created by co-ordinate input and by referencing existing objects

2   Both the WCS and UCS entry modes can be used, but I would recommend:
    a) use the WCS to create the basic model outline
    b) use the UCS to modify and add items to the model

3   It is strongly recommended that a UCS be set and saved for every surface (within reason) on a wire-frame model.

Assignments

Creating wire-frame models at this stage is important as it allows the user to:
a) use 3D co-ordinate entry with the WCS and/or the UCS
b) set and save different UCS positions
c) become familiar with the concept of 3D modelling

I have included three 3D wire-frame models which have to be created. The suggested approach is:

1   Open your 3DSTDA3 standard file – template or drawing

2   Complete the model with layer MODEL current, starting at some convenient point, e.g. 50,50,0. Use WCS entry and add one 'plane' at a time

3   Save each completed model as a drawing file in your named folder with a suitable name, e.g. **C\MODR2004\ACT2**, etc.

4   *Note*:
    a) **do not attempt to add dimensions**
    b) do not attempt to display the two models on 'one screen' – you will soon be able to achieve this for yourself
    c) these models will be used for later assignments, so ensure they are saved
    d) use your discretion for any sizes not given

The activities concern our master builder MACFARAMUS, and you have to create 3D wire-frame models of three of his famous shaped blocks. These blocks were used by MACFARAMUS in other activities, e.g. roads, garden walls, etc.

Activity 2: MACFARAMUS's simple shaped block 1

A relatively simple wire-frame model to create. I suggest that you construct it in a similar manner to the worked example, i.e. create the base (bit of thought needed?) then the vertical planes. When complete, save as **MODR2004\ACT2** as it will be used in a later chapter.

Activity 3: MACFARAMUS's simple shaped block 2

Another simple 3DWFM to create. Draw the base, then copy the base outline to give the top. The vertical edges can then easily be added. When complete, save as ACT3.

Activity 4: MACFARAMUS's complex shaped block 3

This shaped block is slightly more difficult due to the curves. How it is created, I will leave for you to work out. When complete, it should be saved as ACT4.

Chapter 5

The UCS

The UCS is one of the basic 3D draughting 'tools' and it has several commands associated with it. Although it was used in the previous chapter, we will now investigate in more detail:

a) setting a new UCS position
b) moving the UCS
c) the UCS toolbars
d) the UCS dialogue box
e) Orthographic UCSs
f) UCS specific commands

Getting started

1   Open your MODR2004\3DWFM model from the previous chapter. This model has several blue objects and green text displayed. There should also be several saved UCS positions. The model is 'positioned' on the black 'sheet border'.

2   Restore the UCS BASE – probably is current?

3   Layer MODEL current and freeze layer TEXT. Refer to Fig. 5.1 which does not display the black sheet border. This is for clarity only.

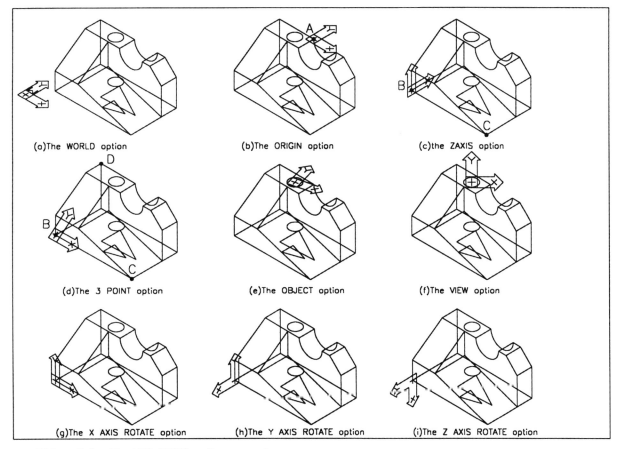

Figure 5.1   The UCS (NEW) options exercise.

Setting a new UCS position

The user can set a new UCS position from the menu bar with **Tools-New UCS** or by entering **UCS <R>** then **N <R>** at the command line. Both methods give the user access to the same options although the selection order differs. The menu bar options are displayed as:

World/Object/Face/View/Origin/Z Axis Vector/3 Point/X/Y/Z

The following exercise is an explanation of these UCS option:

World

1   This option restores the WCS setting irrespective of the current UCS position. It is the default AutoCAD setting.

2   At the command line enter **UCS <R>** then **W <R>** to display the WCS icon on the sheet border at the left vertex as Fig. 5.1(a).

Origin

1   Used to set a new origin point. The user specifies this new origin point by:
    a) picking any point on the screen
    b) co-ordinate entry
    c) referencing existing objects

2   When used, the UCS icon is positioned at the selected point if the UCS Icon display is set to Origin. This option has been used in previous exercises.

3   Menu bar with **Tools-New UCS-Origin** and:
    prompt   Specify new origin point<0,0,0>
    respond   **Intersection icon and pick ptA**
    and   icon positioned as Fig. 5.1(b)

Z Axis Vector

1   Defines the UCS position relative to the Z axis, the user specifying:
    a) the origin point
    b) any point on the Z axis

2   Menu bar with **Tools-New UCS-Z Axis Vector** and:
    prompt   Specify new origin point
    respond   **Intersection icon and pick ptB**
    prompt   Specify point on positive portion of Z axis
    respond   **Intersection icon and pick ptC** – Fig. 5.1(c)

3   The icon will be aligned with:
    a) the X axis along the shorter base left edge
    b) the Y axis along the front left vertical edge
    c) the Z axis along the line BC

3 Point

1   Defines the UCS orientation by specifying three points:
    a) the actual origin point
    b) a point on the positive X axis
    c) a point on the positive Y axis

2 Menu bar with **Tools-New UCS-3 Point** and:
prompt   Specify new origin point
respond **Intersection of ptB**
prompt   Specify point on positive portion of the X axis
respond **Intersection of ptC**
prompt   Specify point on positive-Y portion of the UCS XY plane
respond **Intersection of ptD** – icon as Fig. 5.1(d)

3 This is a very useful option especially if the icon is to be aligned on sloped surfaces. It is probably my preferred method of setting the UCS.

Object

1 Aligns the icon to an object, e.g. a line, circle, polyline, item of text, dimension, block, etc.

2 Menu bar with **Tools-New UCS-Object** and:
prompt Select object to align UCS
respond **pick any point on circle on top surface**

3 The icon is aligned as Fig. 5.1(e) with:
   a) the origin at the circle centre point
   b) the positive X axis pointing towards the circumference of the circle at the point 'picked' by the user.

View

1 Aligns the UCS so that the XY plane is always perpendicular to the view plane

2 Menu bar with **Tools-New UCS-View**

3 The UCS icon will be displayed as Fig. 5.1(f) and is similar to the traditional 2D icon?

4 This is a useful UCS option as it allows 2D text to be added to a 3D drawing – try it for yourself.

X/Y/Z

1 Allows the UCS to be rotated about the entered axis by an amount specified by the user

2 Restore UCS BASE

3 Menu bar with **Tools-New UCS-X** and:
prompt Specify rotation angle about the X axis
enter   **90 <R>** – Fig. 5.1(g)

4 Menu bar with **Tools-New UCS-Y** and:
prompt Specify rotation angle about the Y axis
enter   **−90 <R>** – Fig. 5.1(h)

5 Menu bar with **Tools-New UCS-Z** and:
prompt Specify rotation angle about the Z axis
enter   **−90 <R>** – Fig. 5.1(i).

Face

1   Aligns the UCS with a selected solid model face. This option *cannot be used with 3D wire-frame models*.

2   Restore UCS BASE

3   Menu bar with **Tools-New UCS-Face** and:
   prompt   Select face of solid object
   respond   **pick any line of the top plane**
   prompt   A 3D solid must be selected
               No solids detected

Apply

An option which allows the user to apply the current UCS setting to a specific view-port. We will use this option in later chapters.

Moving a UCS

A selection which allows the user to move the UCS to a new origin position, the UCS icon retaining both its orientation and name. Refer to Fig. 5.1 and:

1   UCS restored to BASE

2   Menu bar with **Tools-Move UCS** and:
   prompt   Specify new origin point or [Zdepth]
   respond   **Intersection icon and pick ptA**
   and   icon moved to point A and retains the name BASE

3   Restore UCS TOP

4   At the command line enter **UCS <R>** and:
   prompt   Enter an option [New/Move/..
   enter   **M <R>** – the move option
   prompt   Specify new origin point or [Zdepth]
   respond   **Intersection icon and pick ptC**
   and   icon moves to point C and retains the name TOP

5   Restore UCS FRONT

6   Menu bar with **Tools-Move UCS** and:
   prompt   Specify new origin point or [Zdepth]
   enter   **Z <R>** – the Z depth option
   prompt   Specify Zdepth<0>
   enter   **−120 <R>**
   and   icon moved to 'back of model' and retains name FRONT

7   *Note*:
   a) This UCS command should be used with caution as the user may not want a named UCS to be 'repositioned'
   b) I never use this command. If I want to reposition the UCS, I use the origin option
   c) Do not save the drawing, as you will save these moved UCSs

8   Task
   Reset the three moved UCSs to their original positions, i.e. BASE, TOP and FRONT.

Other UCS options

The new UCS options are available from the command line but the menu bar selection **Tools-New UCS** is the usual method of activating the command. The command line has other UCS options available for selection, these being:

Prev

1 Restores the previously 'set' UCS position and can be used to restore the last 10 UCS positions.

2 The command is activated from the command line by entering **UCS <R>** then **P <R>** and can be used continually until the command line displays *no previous co-ordinate system saved*.

Restore

1 Allows the user to restore a previously saved UCS position but the names of the saved UCSs must be remembered (this will be modified shortly). This option has been used in our examples.

2 At the command line enter **UCS <R>** then **R <R>** and:
   prompt   Enter name of UCS to restore or [?]
   enter   **TOP <R>**
   then   restore UCS BASE

Save

1 Allows the user to save a UCS position for future recall. It should be used every time a new UCS has been defined.

2 The option is activated from the command line with **UCS <R>** then **S <R>** and the user can enter any name for the UCS position.

Del

1 Entering **UCS <R>** then **D <R>** prompts for the UCS name to be deleted.

2 The default is none. Use with care!

?

1 The query option which will list all saved UCS positions

2 At the command line enter **UCS <R>** then **? <R>** and:
   prompt   Enter UCS name(s) to list <*>
   respond   **press the RETURN key**
   prompt   AutoCAD Text Window with details of the saved UCS co-ordinate systems
   respond   **cancel the window and the command**

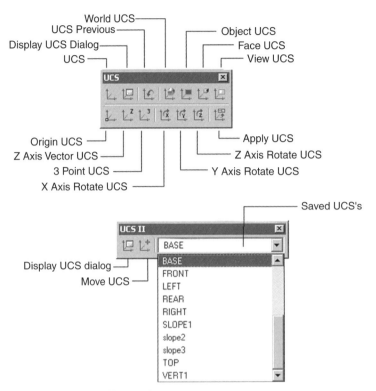

Figure 5.2 The UCS and UCS II toolbars.

The UCS toolbars

All the UCS options have so far been activated by keyboard entry with **UCS <R>** or from the menu bar with **Tools**. The only reason for this is that I think it is easier for the user to understand what option is being used. The UCS options can also be activated in icon form from the UCS and UCS II toolbars – Fig. 5.2. The toolbars have no icon selection for the orthographic options or for Restore, Save, Delete or for query (?), although these can easily be activated by selecting the actual UCS icon. An additional icon in both the UCS and UCS II toolbars is Display UCS Dialog, while the UCS II toolbar allows saved UCSs to be made current, i.e. restored.

The user now has three methods of activating the various UCS options, these being:

a) from the menu bar
b) by command line entry
c) in icon form from the appropriate toolbar

It is user preference as to what method is used.

The UCS dialogue box

The UCS dialogue box can be activated by three different methods:

a) from the menu bar with **Tools-Named UCS**
b) by selecting the Display UCS dialog icon from either the UCS or UCS II toolbar
c) by entering **UCSMAN <R>** at the command line

When activated, the dialogue box allows the user three tab selections, these being:

a) Named UCSs – the default
b) Orthographic UCSs
c) Settings

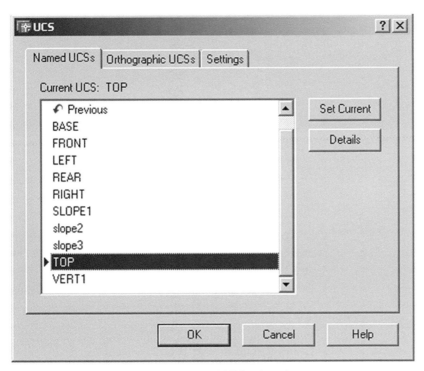

Figure 5.3 The UCS dialogue box – Named UCS tab active.

To demonstrate using the UCS dialogue box:

1 Ensure the 3DWFM is displayed with UCS BASE current

2 Menu bar with **Tools-Named UCS** and:
    prompt   UCS dialogue box
    with   three tab selections and Named UCSs tab active
    and   *a*) a list of saved UCS names for the model
                b) a World and Previous selection option
    respond 1. pick Top – Fig.5.3
             2. pick Set Current
             3. pick OK

2 The model will be displayed with the icon at the TOP setting

3 Use the Named UCS tab of the UCS dialogue box to set current some other saved UCS positions

4 Set UCS BASE current

5 Activate the UCS dialogue box and:
    prompt   UCS dialogue box – Named UCS tab active
    respond 1. pick TOP and it becomes highlighted
             2. right-click the mouse
    prompt   shortcut menu
    with   selections for: Set Current, Rename, Delete, Details
    respond 1. pick Rename
             2. enter new name: **ABOVE <R>**
             3. pick Set Current
             4. pick OK

6 The UCS will be displayed in the 'old top position'

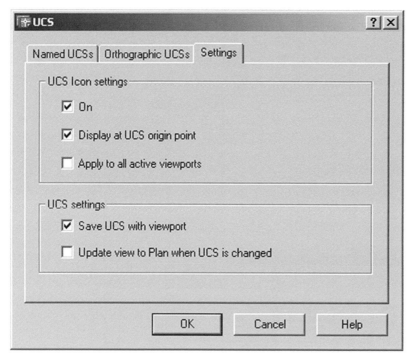

Figure 5.4   The UCS dialogue box – Settings tab active.

7   Now:
  a) rename the ABOVE UCS to TOP again
  b) make UCS BASE current

8   Activate the UCS dialogue box and pick the Settings tab and:
  prompt   Settings tab – Fig. 5.4
  with   1. UCS icon settings for ON and ORIGIN – both active
           2. UCS settings for viewports and plan
  respond **note the settings then pick Cancel**

9   *Note*:
  a) The UCS icon settings from the dialogue box are the same as the menu bar selection of View-Display-UCS Icon-On/Origin
  b) The other Settings options will be discussed in later chapters
  c) The Named tab of the UCS dialogue box can be used to save a UCS position by:
     i) positioning the new UCS as required
     ii) activating the UCS dialogue box, Named UCS tab and there will be an **Unnamed** UCS name listed
     iii) select this Unnamed UCS then right-click
     iv) use the shortcut menu to rename this UCS as required
     v) pick OK
  d) I always use the command line UCS <R> then S <R> to save a new UCS position. I find this easier than the dialogue box method.

Setting an orthographic UCS

This allows the user to restore six preset UCS positions, the orientation being set relative to a saved UCS. Refer to Fig. 5.5 and:

1   Ensure the 3DWFM is displayed with the saved UCSs

2   Restore UCS SLOPE1 current – Fig. 5.5(a)

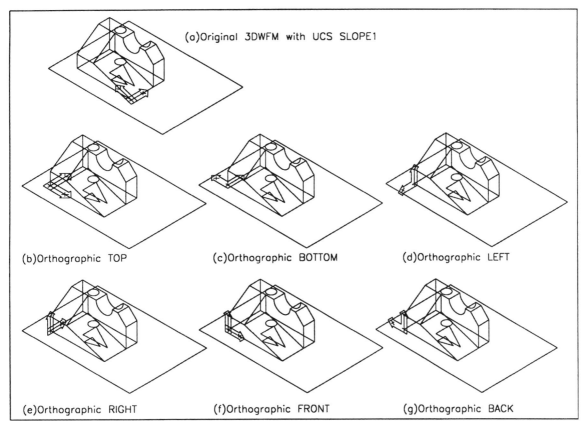

(a)Original 3DWFM with UCS SLOPE1

(b)Orthographic TOP

(c)Orthographic BOTTOM

(d)Orthographic LEFT

(e)Orthographic RIGHT

(f)Orthographic FRONT

(g)Orthographic BACK

Figure 5.5 The Orthographic UCS options exercise relative to BASE with UCS SLOPE1 current.

3 Activate the UCS dialogue box with the Orthographic UCS tab active and:

   prompt   UCS dialogue box – Orthographic tab display

   respond 1. scroll at Relative to
                2. pick BASE
                3. at Current UCS Name, pick TOP
                4. pick Set Current – Fig. 5.6
                5. pick OK

   and   icon displayed as Fig. 5.5(b)

4 With the Orthographic tab of the UCS dialogue box, set Bottom current relative to BASE – Fig. 5.5(c)

5 Menu bar with **Tools-Orthographic UCS-Left** to display the icon as Fig. 5.5(d)

6 At the command line enter **UCS <R>** and:

   prompt   Enter an option [New/Move/orthoGraphic/..

   enter   **G <R>** – the orthographic option

   prompt   Enter an option [Top/Bottom/Front/..

   enter   **R <R>** – the right orthographic UCS option

   and   the icon will be displayed as Fig. 5.5(e)

7 With the menu bar **Tools-Orthographic UCS** sequence, select:

   a) FRONT current – Fig. 5.5(f)

   b) BACK current – Fig. 5.5(g)

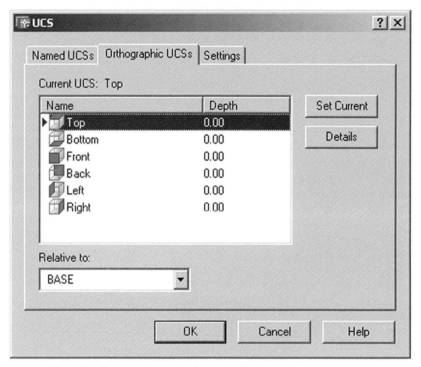

Figure 5.6   The UCS dialogue box – Orthographic tab active.

8 Activate the Orthographic tab of the UCS dialogue box and:
   a) set relative to SLOPE1
   b) activate the six orthographic UCS positions
   c) note the orientation of the UCS with each orthographic name
   d) restore UCS BASE

9 This exercise with the UCS dialogue box is now complete.

10 *Note*:
   The menu bar sequence **Tools-Orthographic UCS-Presets** will display the Orthographic tab of the UCS dialogue box.

UCS specific commands

The UCS has several specific commands associated with it. At this stage, we will only investigate:
a) the PLAN command
b) the UCSFOLLOW system variable.

Plan

Plan is a command which displays any model perpendicular to the XY plane of the current UCS position.

1 Ensure the 3DWFM is displayed with UCS BASE current

2 Refer to Fig. 5.7

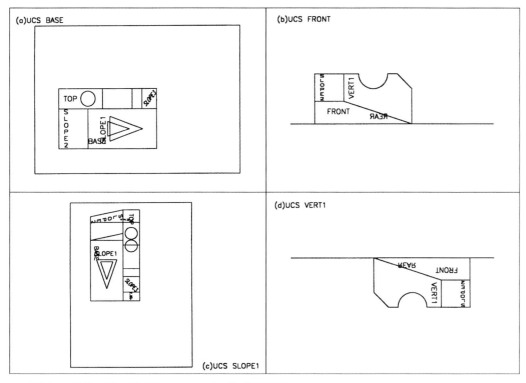

Figure 5.7   The PLAN command with 3DWFM.

3   At the command line enter **PLAN <R>** and:
    prompt   Enter an option [Current ucs/Ucs/World] <Current>
    enter   **<R>** i.e. accept the Current UCS default

4   The screen will display the model as a plan view – Fig. 5.7(a). This view is perpendicular to the current UCS setting (BASE) and is really a 'top' view in orthogonal terms

5   Restore UCS FRONT – pencil icon displayed?

6   Menu bar with **View-3D Views-Plan View-Current UCS** and the model will be displayed as Fig. 5.7(b). This is a plan view to the current UCS FRONT and is a 'front' view in orthogonal terms.

7   Menu bar with **View-3D Views-Plan View-Named UCS** and:
    prompt   Enter name of UCS
    enter   **SLOPE1 <R>**

8   The model will be displayed as a plan to the UCS SLOPE1 setting as Fig. 5.7(c)

9   At the command line enter **PLAN <R>** and:
    prompt   Enter an option [Current ucs/Ucs/World]<Current>
    enter   **U <R>** – the Ucs option
    prompt   Enter name of UCS
    enter   **VERT1 <R>**

10 The model display is as Fig. 5.7(d), i.e. a plan view to the UCS setting VERT1. This display should be upside-down – why?

11 Finally restore UCS BASE and menu bar with View-3D Views-SE Isometric to return the original model display.

UCSFOLLOW

UCSFOLLOW is a system variable which controls the screen display of a model when the UCS position is altered. The variable can only have the values of 0 (default) or 1 and:
a) UCSFOLLOW 0: no effect on the display with UCS changes
b) UCSFOLLOW 1: automatically generates a plan view when the UCS is altered

1   The 3DWFM model still displayed with UCS BASE?

2   At the command line enter **UCSFOLLOW <R>** and:
    prompt   Enter new value for UCSFOLLOW <0>
    enter   **1 <R>**

3   Nothing has changed?

4   Restore UCS FRONT – plan view as Fig. 5.7(b)

5   Restore UCS SLOPE1 – plan view as Fig. 5.7(c)

6   Restore UCS VERT1 – plan view as Fig. 5.7(d)

7   Restore UCS BASE – plan view as Fig. 5.7(a)

8   Set UCSFOLLOW back to 0 and restore the original screen display with View-3D Views-SE Isometric

9   This completes the exercises with the UCS.

Summary

1   The UCS is an essential 3D modelling aid

2   The UCS command has several options including:
    a) New: origin, 3 point, X,Y,Z rotate
    b) Move: which should be used with caution
    c) Orthographic: six preset UCS settings

3   The orientation of the UCS icon is dependent on the option used

4   The UCS toolbars offer fast option selection

5   It is **STRONGLY RECOMMENDED** that the UCS icon and the UCS icon origin are ON when working in 3D. These can be activated with:
    a) the menu bar sequence View-Display-UCS Icon
    b) the Settings tab of the UCS dialogue box

6   The UCS dialogue box allows flexible management of the UCS with three tab selections:
    a) Named UCSs – set current, rename, delete
    b) Orthographic UCSs (Presets)
    c) Settings

7   PLAN is a command which displays the model perpendicular to the XY plane of the current UCS

8   UCSFOLLOW is a system variable which can be set to give automatic plan views when the UCS is repositioned. It is recommended that this variable be set to 0, i.e. off.

Chapter 6

The modify commands with 3D models

All the modify commands are available for use with 3D models, but the results are dependent on the UCS position. We will investigate how the COPY and ARRAY commands can be used with our 3D wire-frame model so:

1   Open your 3DWFM model with UCS BASE and layer MODEL current
2   Display the Modify, Object snap and UCS toolbars
3   Erase all text except the FRONT text item.

The COPY command

1   Select the COPY icon from the Modify toolbar and:

prompt   Select objects
respond **pick the 4 red lines and the green FRONT text item on the 'front vertical' plane then right-click**
prompt   Specify base point or displacement
respond **Intersection icon and pick ptA**
prompt   Specify second point of displacement
enter   **@0,0,260 <R>** – Fig. 6.1.A(a)

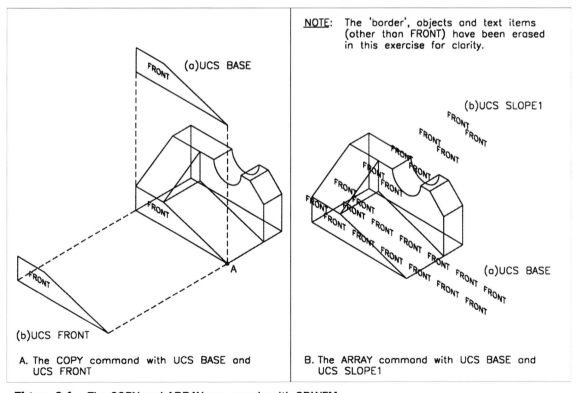

NOTE: The 'border', objects and text items (other than FRONT) have been erased in this exercise for clarity.

(a)UCS BASE

(b)UCS SLOPE1

(a)UCS BASE

(b)UCS FRONT

A. The COPY command with UCS BASE and UCS FRONT

B. The ARRAY command with UCS BASE and UCS SLOPE1

Figure 6.1   The COPY and ARRAY commands with 3DWFM.

2 Restore UCS FRONT

3 Select the COPY icon and:
   prompt   Select objects
   respond **pick the same 5 objects as before then right-click**
   prompt   Specify base point
   respond **pick Intersection of ptA**
   prompt   Specify second point
   enter   **@0,0,260 <R>** – Fig. 6.1.A(b)

4 Menu bar with **View-Zoom-All**

5 Undo (or erase) the copied effects to leave the original model.

The ARRAY command

1 Restore UCS BASE

2 Menu bar with **Modify-Array** and using the array dialogue box:
   a) Select objects: pick the FRONT text item then right-click
   b) Rows: 2
   c) Columns: 6
   d) Offsets: Row 40, Column 60
   e) Angle of array: 0 – Fig. 6.2
   f) Pick Preview then Accept

3 The text item will be arrayed in a 2 × 6 rectangular matrix as Fig. 6.1.B(a)

4 Restore UCS SLOPE1

5 Rectangular array the original FRONT text item using the same entries as step 2 – Fig. 6.1.B(b)

6 Undo (or erase) the arrayed effects.

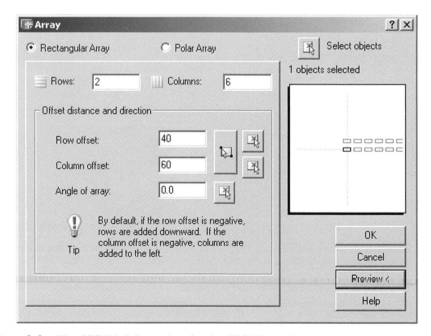

Figure 6.2   The ARRAY dialogue box for the FRONT text item.

Other modify commands

Although only two of the modify commands have been demonstrated in this chapter, all of the modify commands are available for use with 3D models, but the final result is dependent on the UCS position. The only requirement for the user is to ensure that the icon is positioned 'correctly' for the modification.

Chapter 7

Dimensioning in 3D

There are no special commands to add dimensions in 3D. Dimensioning is a 2D concept, the user adding the dimensions to the XY plane of the current UCS setting. This means that the orientation of the complete 'dimension object' will depend on the UCS position. The user should be aware of:

a) AutoCAD's automatic dimensioning facility

b) linear dimensioning will be horizontal or vertical, depending on where the dimension line is located in relation to the object being dimensioned.

We will demonstrate how dimensions can be added to 3D models with two examples. The first will be the 3DWFM, and the second will use AutoCAD's stored 3D objects.

Example 1

1 Open MODR2004\3DWFM and display the Dimension, Object Snap and other toolbars to suit

2 Freeze layer TEXT and make layer DIM current

3 The standard sheet created as the template/drawing file had a created dimension style – 3DSTD. You may want to 'alter' the Overall Scale (Fit tab) to a value of 1.5 which will make the added dimensions 'clearer'.

4 Ensure UCS BASE is current and refer to Fig. 7.1

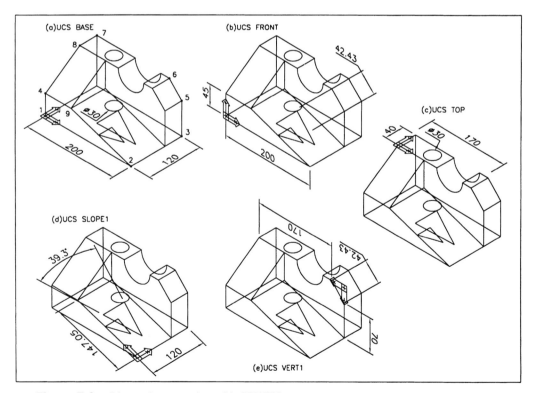

Figure 7.1   Dimension exercise with 3DWFM.

5 Select the LINEAR dimension icon and:
   prompt   Specify first extension line origin
   respond **Intersection icon and pick pt1**
   prompt   Specify second extension line origin
   respond **Intersection icon and pick pt2**
   prompt   Specify dimension line location
   respond **pick to suit**

6 Repeat the LINEAR dimension selection and dimension line 23, positioning the dimension line to suit

7 Select the DIAMETER icon and:
   prompt   Select circle or arc
   respond **pick the circle on the TOP 'plane'**
   prompt   Specify dimension line location
   respond **pick to suit** – interesting result?

8 The three added dimensions will be displayed as Fig. 7.1(a)

9 Erase the added dimensions and restore UCS FRONT

10 Using the dimension icons:
   a) linear dimension lines 12 and 14
   b) align dimension line 56
   c) try and add a diameter dimension to the top circle
   d) dimensions displayed as Fig. 7.1(b)

11 Erase these added dimensions and restore UCS TOP and:
   a) linear dimension line 67 and line 78
   b) diameter dimension the circle on the top
   c) result as Fig. 7.1(c)

12 Restore UCS SLOPE1, erase the previous dimensions and:
   a) linear dimension line 23 and line 29
   b) angular dimension a vertex of the blue triangle 'on the slope'
   c) the three dimensions will be displayed as Fig. 7.1(d)

13 With UCS VERT1 current, erase the dimensions from SLOPE1 and:
   a) linear dimension line 67 and line 35
   b) align dimension line 56
   c) interesting result as Fig. 7.1(e) – why?

14 This exercise should demonstrate to the user that:
   a) adding dimensions to a 3D model is **VERY UCS DEPENDENT**
   b) there are no special 3D commands
   c) the actual orientation of added dimensions depends on the UCS position and orientation
   d) dimensions are added to the XY plane of the current UCS

15 *Task*
   a) erase any dimensions still displayed
   b) with layer DIM current refer to Fig. 7.2 and add the given dimensions to the model
   c) some of the existing saved UCS positions will be used
   d) you may have to set a new UCS position for the continuous 80,40 and the 70 dimensions
   e) when complete save if required, but not as 3DWFM

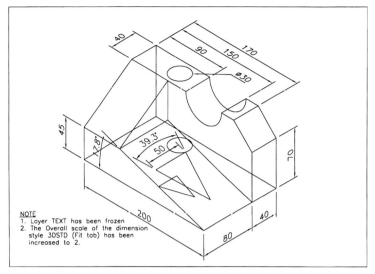

Figure 7.2   Required dimensions to be added to 3DWFM.

Example 2

1 Close any existing drawings then menu bar with **File-New** and:
   a) pick Use a Template
   b) pick **3DSTDA3.dwt** (your 3D template file) then pick OK

2 The screen will display the black border:
   a) in SE Isometric viewpoint
   b) layer MODEL current
   c) WCS icon at left vertex of border

3 Display the Surfaces toolbar and refer to Fig. 7.3

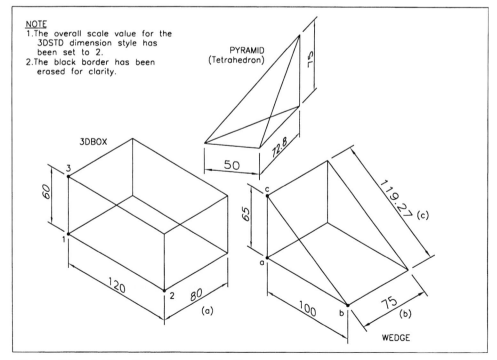

Figure 7.3   Adding dimensions to 3D objects.

4 Menu bar with **Draw-Surfaces-3D Surfaces** and:
  prompt `3D Objects dialogue box`
  respond **pick Box3d** then OK
  prompt `Specify corner of box and enter:` **50,50,0 <R>**
  prompt `Specify length of box and enter:` **120 <R>**
  prompt `Specify width of box or [Cube] and enter:` **80 <R>**
  prompt `Specify height of box and enter:` **60 <R>**
  prompt `Specify rotation of box about Z axis and enter:` **0 <R>**

5 Select the WEDGE icon from the surface toolbar and:
  prompt `Specify corner of wedge and enter:` **200,150,0 <R>**
  prompt `Specify length of box and enter:` **100 <R>**
  prompt `Specify width of box and enter:` **75 <R>**
  prompt `Specify height of box and enter:` **65 <R>**
  prompt `Specify rotation of wedge about Z axis and enter:` **0 <R>**

6 Using the 3 point UCS option:
  a) origin at point 1
  b) X axis along line 12
  c) Y axis along line 13
  d) save UCS position as POS1
  e) add two linear dimensions with this UCS

7 Use the 3 point UCS option again with:
  a) origin at point a
  b) X axis along line ab
  c) Y axis along line ac
  d) save as POS2
  e) add two linear dimensions with this UCS

8 Now add the three other linear dimensions (a), (b) and (c). Some UCS manipulation is required for this.

9 Menu bar with **Tools-New UCS-World**

10 *Task*
  Another 3D object has to be created and dimensioned so select the pyramid icon from the Surfaces toolbar and:
  prompt `Specify first corner point for base of pyramid`
  enter **40,220 <R>**
  prompt `Specify second corner point for base of pyramid`
  enter **80,250 <R>**
  prompt `Specify third corner point for base of pyramid`
  enter **60,320 <R>**
  prompt `Specify fourth corner point for base of pyramid or [Tetrahedron]`
  enter **T <R>** – the tetrahedron option
  prompt `Specify apex point of tetrahedron`
  enter **60,320,75 <R>**

11 Now set appropriate UCS positions and add the three linear dimensions as displayed

12 Save if required, but the drawing will not be used again

Summary

1 There are no special 3D commands

2 Dimensioning is a 2D concept and dimensioning a 3D model involves adding the dimensions to the XY plane of the required UCS setting

3 If the UCS is not positioned correctly, dimensions can have the 'wrong orientation' in relation to the object being dimensioned.

Assignment

Two dimensioning assignments for you to complete, one being an existing wire-frame model, the other being a new wire-frame model. Both involve the work of our great architect MACFARAMUS.

Activity 5: Shaped block (No 1) of MACFARAMUS.

Recall the shaped block from *Activity* 2 and:
a) set and save appropriate UCS positions
b) add the given dimensions
c) save the completed exercise for later recall.

Activity 6: The famous rectangular topped pyramid of MACFARAMUS.

The rectangular topped pyramid designed by MACFARAMUS was never built but his design is still considered unique. You have to:
a) create a wire-frame model of the pyramid using the sizes given
b) set and save several UCS positions, these being on the following surfaces:
    1. the base and top – UCS BASE and UCS TOP
    2. the four 'sides' of the pyramid, which I have named as SLOPE1, SLOPE2, SLOPE3 and SLOPE4 (I know that two of these sides are not sloped but I named them in a logical order.)
    3. the vertical sides of the top. Use your own names, e.g. V1, etc.
c) add the given dimensions
d) save the completed model

MACFARAMUS had to calculate certain angles and distances without the aid of calculators or computers. This took him a great deal of time and involved using his knowledge of trigonometry. The angles and distances he required were:
1. the angle of the two sloped 'planes' in relation to the base
2. the length of the three sloped 'edges'

His calculations gave the following values:
a) angle of first sloped plane to the base: 50.19
b) angle of second sloped plane to the base: 56.31
c) length of the three sloped edges:
    1 – 234.31
    2 – 263.25
    3 – 216.33

Was he correct?

Chapter 8

Hatching in 3D

There are no special 3D hatch commands. Hatching (like dimensioning) is a 2D concept, the hatch pattern being added to the XY plane of the current UCS. Two examples will be used to demonstrate adding hatching to 3D models.

Example 1

1 Open your 3DSTDA3 template file
2 Display the Draw, Modify and Object Snap toolbars
3 With layer MODEL current, refer to Fig. 8.1 and draw the four 'mutually perpendicular planes' using the LINE icon with:

	Plane 1234	*Plane 1564*	*Plane 3764*	*Plane 7896*
Start	30,30,0	30,30,0	130,130,0	130,130,100
Next	@100,0,0	@0,0,100	@0,0,100	@0,100,0
Next	@0,100,0	@0,100,0	@−100,0,0	@−100,0,0
Next	@−100,0,0	@0,0,−100	<RETURN>	@0,−100,0
Next	close	<RETURN>		<RETURN>

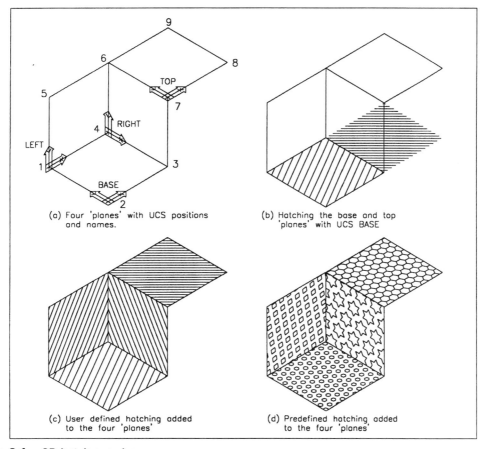

(a) Four 'planes' with UCS positions and names.

(b) Hatching the base and top 'planes' with UCS BASE

(c) User defined hatching added to the four 'planes'

(d) Predefined hatching added to the four 'planes'

Figure 8.1 3D hatch exercise.

4 Erase the black border and pan the drawing to suit

5 Set and save the four UCS positions as Fig. 8.1(a)

6 Restore UCS BASE and make SECT the current layer

7 Select the HATCH icon from the Draw toolbar and using the Boundary Hatch dialogue box with the Hatch tab active:
   a) select User defined pattern type
   b) set angle to 30 and spacing to 8
   c) use the **Pick Points<** option and:
      1. select a point within the 1234 plane then right-click/enter
      2. Preview then right-click to accept the hatching.

8 Repeat the HATCH icon selection and:
   a) using the Pick Points option pick a point within the 6789 plane and:
      prompt   Boundary Definition Error dialogue box – Fig. 8.2
      respond   **pick OK then <RETURN>**
   b) using the Select Objects option pick the four lines of the 6789 plane then right-click/enter and:
      1. set angle to −45 and spacing to 5
      2. Preview then right-click to accept the hatching.

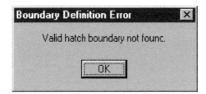

Figure 8.2   The Boundary Definition Error Message box.

9 The result of the two hatch operations is displayed in Fig. 8.1(b) with plane 1234 having the correct hatching, but plane 6789 has none, the hatching having been added to the plane of UCS BASE.

10 Use the HATCH icon and try and add hatching to the vertical planes 1564 and 3467.
   a) Not possible?
   b) Unable to hatch the boundary message at the command line
   c) Why is this?

11 Erase the 'wrong' hatching added in step 8 then restore UCS TOP

12 Hatch the top plane (6789) using the HATCH icon with:
   a) pick points option
   b) angle −45 and spacing 5

13 Add hatching to the two vertical planes remembering to restore UCS LEFT and UCS RIGHT – Fig. 8.1(c). Use your own angle and spacing values.

14 *Task*
   a) Erase the added hatching then add the following predefined hatch patterns using the information given:

UCS	Pattern	Scale	Angle
BASE	HEX	1	−10
TOP	HONEY	2	0
LEFT	SQUARE	2	10
RIGHT	STARS	2	0

   c) the result is Fig. 8.1(d)
   d) this completes Example 1 which does not have to be saved.

Example 2

1   Open C:\MODR2004\3DWFM and refer to Fig. 8.3

2   *a*) erase all text and any dimensions
     b) erase the smaller triangle and lower circle on the base plane
     c) make layer SECT current
     d) restore UCS TOP

3   Select the HATCH icon and:
     a) Predefined pattern type: scroll and pick STARS
     b) set scale to 1 and angle to 0
     c) select Pick Points: pick internal points within the **TWO** top planes then right-click/enter
     d) preview and right-click

4   Restore UCS FRONT and with the HATCH icon:
     a) pattern type: ZIGZAG
     b) scale: 1.5 and angle: 30
     c) select objects: pick the four lines of front vertical plane then right-click/enter
     d) preview and right-click

5   Repeat the HATCH icon selection and add the following hatch patterns:

UCS	*Pattern*	*Scale*	*Angle*	*Selection type*
SLOPE1	TRIANG	1	25	pick points
VERT1	HOUND	2.5	0	pick points

6   Menu bar with **View-Hide** and the model is as before. Hatching a wire-frame model does not produce a hide effect.

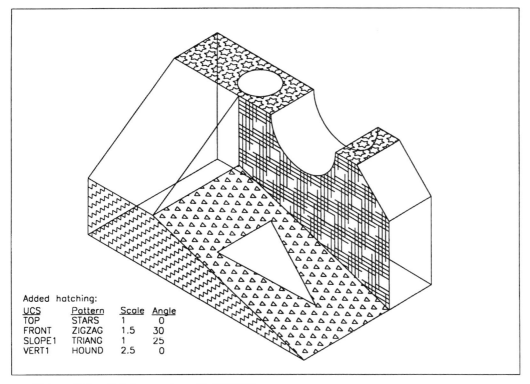

Added hatching:

UCS	Pattern	Scale	Angle
TOP	STARS	1	0
FRONT	ZIGZAG	1.5	30
SLOPE1	TRIANG	1	25
VERT1	HOUND	2.5	0

Figure 8.3   Hatch exercise with 3DWFM.

7   Save your completed hatched model but not as MODR2004/3DWFM.

8   *Note*:
     In the two hatch exercises we used the one layer (SECT) for all hatching. It is some-
     times desirable to have a different layer for each current UCS that is to be used for
     hatching. This is a user decision.

Summary

1   Hatching is a 2D concept

2   Hatching a 3D model requires the UCS to be set to the 'plane' which is to be hatched

3   *a*) both user-defined and predefined hatch patterns can be used
     b) both the select objects and pick points methods are permitted

4   Hatching a 3D wire-frame model does not produce a hide effect

Assignment

A single hatch activity for you to attempt.

Activity 7: The rectangular topped pyramid of MACFARAMUS.

1   Open the model drawing

2   Freeze layer DIM

3   Using the correct UCS, add the following predefined hatch patterns:

UCS	Pattern	Scale	Angle
four 'base' planes	BRICK	2	0
four vert 'top' planes	BRSTONE	1	0
the horiz top surface	EARTH	1.5	0

4   Do you make a new hatch layers for each UCS position?
     Your decision!

5   Decide on whether to use the pick points or select objects option for adding the hatching

6   Save the completed model as **MODR2004/PYRAMID**

7   I have displayed the activity model at two 3D viewpoints for effect.

Chapter 9

Tiled viewports

1 Up until now, only one view of created models has been displayed with the Model tab active (see the Status bar). The graphics screen can however, be divided into a number of separate viewing areas called **viewports** and each viewport can display different viewpoints of a model.

2 Viewports are **interactive**, i.e. what is drawn in one viewport is automatically drawn in the other viewports and the user can switch between viewports when creating a model. Viewport layouts (**configurations**) can be saved thus allowing different displays of the same model to be stored for future recall.

3 Viewports are essential with 3D modelling as they allow different views of the model to be displayed on the screen simultaneously

4 When used with the VIEWPOINT command (next chapter) the user has a very powerful 3D draughting tool

5 There are two types of viewport (displayed in Fig. 9.1) these being:
   a) Tiled or fixed
   b) Untiled or floating

6 The type of viewport which is displayed is controlled by the **TILEMODE** system variable and:
   a) Tilemode 1: tiled viewports – the default setting. These viewports are fixed and cannot be altered by the user.
   b) Tilemode 0: untiled viewports – can be altered by the user.

7 In this chapter we will only investigate TILED viewports and leave the untiled viewport discussion to a later chapter when we will investigate model and paper space

8 Making viewports can be activated by keyboard entry or from the menu bar.

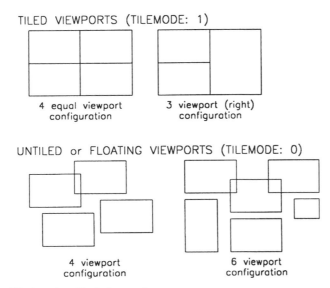

Figure 9.1   Tiled and untiled viewports.

Example 1

This exercise is rather long, but persevere with it.

1 Open your 3DWFM drawing of the wire-frame model on the black border with layer MODEL current and UCS BASE

2 Deactivate all floating toolbars and display the model without any text, dimensions or hatching. Erase or freeze layers?

3 At the command line enter **TILEMODE <R>** and:
   prompt   Enter new value for TILEMODE<1>
   and   observe the 1 default then press **ESC**

4 The TILEMODE value of 1 indicates that only TILED viewports can be used. The same condition is also evident with:
   a) Status bar: word MODEL is displayed
   b) Menu bar: View-Viewports indicate that Polygonal Viewport and Object are not available for selection (these are paper space options)

5 Menu bar with **View-Viewports-3 Viewports** and:
   prompt   Enter a configuration [Horizontal/Vertical/Above..
   enter   **R <R>** – the right configuration option

6 The drawing screen will:
   a) be divided into three separate 'areas' – one large at the right and two smaller to the left. The three viewports will 'fill the screen' as Fig. 9.2(a)
   b) display the same view of the model in the three viewports

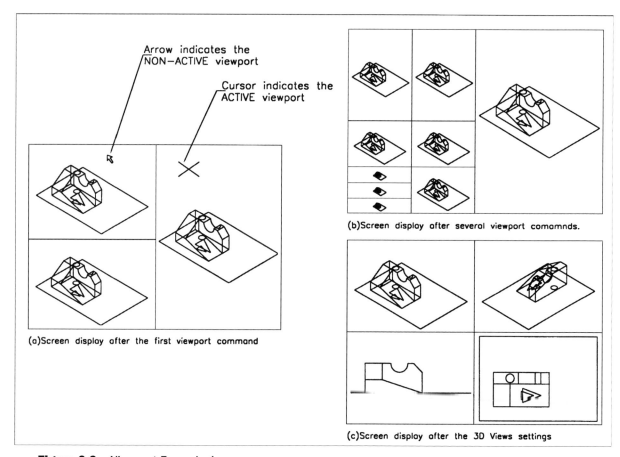

(a)Screen display after the first viewport command

(b)Screen display after several viewport comamnds.

(c)Screen display after the 3D Views settings

Figure 9.2   Viewport Example 1.

7 Move the mouse about the screen and:
   a) the large viewport will display the cursor cross-hairs and is the **ACTIVE** viewport, i.e. it is 'current'
   b) the other viewports will display an arrow and these viewports are **NON-ACTIVE**

8 Any viewport can be made active by:
   a) moving the mouse into the viewport area
   b) left-click
   c) try this for yourself a few times

9 At the command line enter **-VPORTS <R>** and:
   prompt   Enter an option [Save/Restore/Delete/..
   enter   **S <R>** – the save option
   prompt   Enter name for new viewport configuration
   enter   **CONF1 <R>**

10 Make the upper left viewport active and select the menu bar sequence **View-Viewports-2 Viewports** and:
   prompt   Enter a configuration option [Horizontal/Vertical..
   enter   **V <R>** – the vertical option
   and   the top left viewport will be further divided into two equal vertical viewports, each displaying the model layout

11 Make the lower left viewport active and menu bar with **View-Viewports-4 Viewports** to display an additional four viewports of the model

12 At the command line enter **-VPORTS <R>** and:
   prompt   Enter an option [Save/Restore/..
   enter   **S <R>** – the save option
   prompt   Enter name for new viewport configuration
   enter   **CONF2 <R>**

13 With the lower left viewport active, enter **-VPORTS <R>** at the command line and:
   prompt   Enter an option [Save/Restore..
   enter   **3 <R>** – the 3 viewport option
   prompt   Enter a configuration option [Horizontal/Vertical/Above..
   enter   **H <R>** – the horizontal option

14 The lower left viewport will be further divided into another three viewports and at this stage your screen should resemble Fig. 9.2(b)

15 Make the lower left viewport active and enter **-VPORTS <R>** then **4 <R>** and the following message will be displayed at the prompt line: *The current viewport is too small to divide*

16 Save the screen viewport configuration as CONF3 – easy?

17 *a*) make the large right viewport active
   b) menu bar with View-Viewports-1 Viewport
   c) original screen displayed?
   d) zoom-all needed?

18 Menu bar with **View-Viewports-4 Viewports** to 'fill the screen' with four viewports of the model

19 Using the menu bar **View-3D Views** selection make each viewport current and set different viewpoints using the following:
Viewport	*3D View*
top left	SE Isometric
top right	NE Isometric
lower right	Plan-Current UCS
lower left	Front

20 The screen display should resemble Fig. 9.2(c)

21 Save the screen configuration as CONF4

22 *Task*
    Restore the screen to a single viewport configuration to display the original model layout

23 At the command line enter **-VPORTS <R>** and:
    prompt   Enter an option [Save/Restore..
    enter   **R <R>** – the restore option
    prompt   Enter name of viewport configuration to restore
    enter   **CONF1 <R>**
    and   screen displays the first saved configuration

24 Restore the other three saved viewport configurations using the command line–VPORTS, then restore the display to a single viewport

25 *Notes*:
    a) The command line entry **-VPORTS** gives the user the viewport options at the command line. This was deliberate for this first example.
    b) Generally the viewports command is activated from the menu bar in dialogue box form

26 Menu bar with **View-Viewports-Named Viewports** and:
    prompt   Viewports dialogue box
    with   Named Viewports tab active
    and   four saved viewport configurations
    respond **pick CONF3 then OK** – Fig. 9.3

27 The screen will display the named viewport configuration

28 Using the Named viewport dialogue box, display the other named viewports then restore the model in the original single viewport as opened

29 This completes the first viewport exercise. If you want to save the exercise (with the viewport configurations) **DO NOT USE THE NAME 3DWFM**

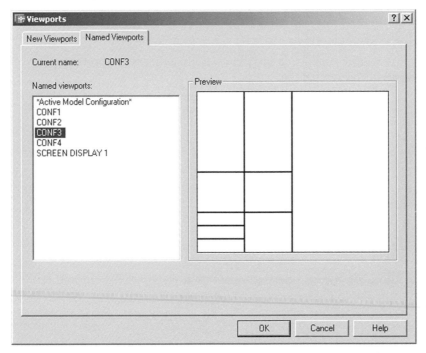

Figure 9.3   The Viewports dialogue box with the Named Viewports tab active.

Example 2

The first exercise used an already created 3D model to investigate the viewport command and configurations. This current exercise will create a new 3D wire-frame model interactively using a four viewport configuration with preset 3D viewpoints. This will allow the user to 'see' the model being created in all four viewports at the one time.

1 Open your **3DSTDA3** template file to display the black border at a 3D viewpoint with layer MODEL current

2 Menu bar with **View-Display-UCS Icon** and check both On and Origin are active (tick) – they should be!

3 Menu bar with **Tools-New UCS-Origin** and:
   prompt   Specify new origin point
   enter   **50,50,0 <R>**
   and   icon moves to the entered point and is displayed as a UCS icon

4 Save this UCS position as BASE

5 Menu bar with **View-Viewports-New Viewports** and:
   prompt   Viewports dialogue box with New Viewports tab active
   respond 1. New name: enter SCREEN DISPLAY 1
             2. Standard viewports: pick Four: Equal
             3. Apply to: Display
             4. Setup: scroll and pick 3D
             5. Change view to: do not alter (Fig. 9.4)
             6. pick OK

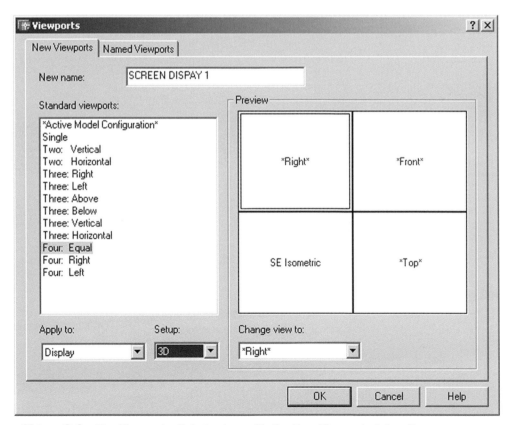

Figure 9.4   The Viewports dialogue box with the New Viewports tab active.

6 The screen will display a four viewport configuration with the black border displayed in each. Note the 'appearance' of the icon in the top two, and lower right viewports – it has the same configuration in each, despite the different viewpoints set in the New Viewports dialogue box (respond 4 in step 5).

7 Making each viewport active in turn, enter the following at the command line **UCSVP <R>** and:
   prompt Enter new value for UCSVP<1>
   enter   **0 <R>**

8 Making each viewport active in turn, at the command line enter **ZOOM <R>** then **0.9 <R>**

9 The screen layout at this stage is similar to Fig. 9.5(a)

10 With the lower left viewport active, construct the model base using the LINE icon with:
   Start point **0,0,0 <R>**   pt1
   Next point   **@200,0,0 <R>**   pt2
   Next point   **@0,120,0 <R>**   pt3
   Next point   **@200<180,0 <R>**   pt4
   Next point   **close** – Fig. 9.5(b) in 3D

11 Using the LINE command construct the front vertical side with:
   Start point **Endpoint of pt1**
   Next point   **@20,0,100 <R>**   pt5
   Next point   **@120,0,0 <R>**   pt6
   Next point   **Endpoint of pt2**
   Next point   **right-click/enter** – Fig. 9.5(c) in 3D

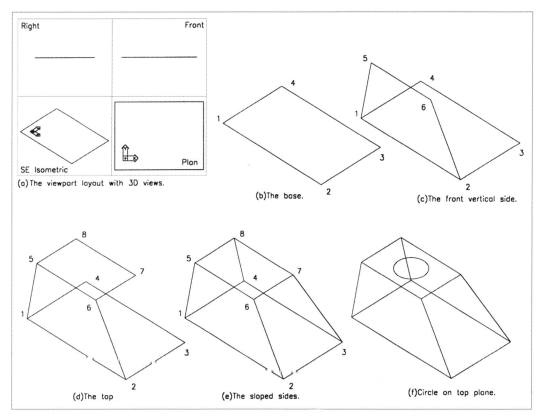

Figure 9.5 Construction of model for viewport Example 2.

12 The top surface is created with the LINE command and:
Start point **Endpoint of pt6**
Next point **@0,80,0 <R>** pt7
Next point **@−120,0,0 <R>** pt8
Next point **Endpoint of pt5**
Next point **right-click/enter** – Fig. 9.5(d) in 3D

13 Add the sloped sides with lines joining points 3–7 and 4–8 as Fig. 9.5(e) in 3D

14 Make layer OBJECTS (blue) current and draw a circle with centre: 80,40,100 and radius: 25 – Fig. 9.5(f) in 3D

15 Menu bar with **Draw-Surfaces-3D Surfaces** and:
prompt 3D Objects dialogue box
respond **pick Box3d then OK**
prompt Specify corner of box and enter: **80,30,0**
prompt Specify length of box and enter: **50**
prompt Specify width of box and enter: **40**
prompt Specify height of box and enter: **30**
prompt Specify rotation angle of box about Z axis and enter: **20**

16 *a*) Make layer TEXT current
   b) Rotate UCS about X axis by 90 and save as FRONT
   c) Menu bar with **Draw-Text-Single Line Text** and add the text item AutoCAD, **centred** on 80,50 with height 20 and rotation 0

17 *a*) Set a 3 point UCS on the right sloped surface with:
      1. origin: midpoint of line 23 – Fig. 9.5(e)
      2. x axis: intersection of pt3
      3. y axis: perpendicular to line 67
      4. save UCS as SLOPE
   b) Add the single line text item R2004, **centred** on −5,50 with a height of 15 and a rotation angle of 0

18 The complete four viewport configuration display should be similar to Fig. 9.6

19 Save the drawing as **MODR2004\TEST3D**

20 This completes the two exercises on viewports

21 *Notes*:
   1. A new system variable was used during this exercise, this being **UCSVP**. This variable determines whether the UCS in an active viewport will 'reflect' the UCS orientation of that active viewport and:
      a) UCSVP 0: unlocked, i.e. the UCS will reflect the UCS of the current active viewport
      b) UCSVP 1: locked, i.e. UCS is independent of the UCS in the current active viewport
   2. The default UCSVP value is 1, i.e. locked
   3. It is **my personal recommendation that the UCSVP is set to 0**, i.e. it should always reflect the UCS position in any active viewport
   4. The UCSVP must be set in every created viewport.

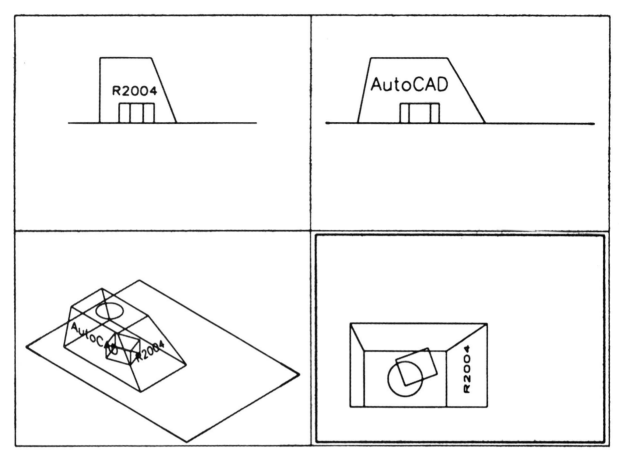

Figure 9.6 Completed viewport example 2 – TEST3D.

Summary

1 Viewports allow multi-screen configurations to be set

2 There are two types of viewport – TILED and UNTILED

3 The viewport type is controlled by the system variable TILEMODE and:
   a) TILEMODE 1: tiled viewports (fixed)
   b) TILEMODE 0: untiled viewports (movable) – more later

4 Tiled viewports can have between 1 and 4 'divisions' and 'fill the screen drawing area'

5 Multi-screen viewports are generally used with the viewpoint command and their full benefit will not be appreciated until the various viewpoint options are discussed

6 Multiple viewport layouts are essential to 3D modelling.

Chapter 10

3D Views (Viewpoints)

3D Views (or viewpoints) determine how the user 'looks' at a model and has been used in previous chapters without any discussion about how it is used. In this chapter we will investigate the command in detail using previously created models. When combined with viewports, the user has a very powerful draughting aid – multiple viewports displaying different views of a model.

The viewpoint command has the following selection options:
a) Isometric views: SW, SE, NE, NW
b) Orthographic views: Top, Bottom, Left, Right, Front, Back
c) Plan view: to current UCS, WCS, named UCS
d) Viewpoint: with rotate, compass and tripod, vector options
e) Viewpoint Presets: dialogue box selection
f) Real-time rotation with 3D Orbit
g) New Viewports dialogue box

In this chapter we will investigate all of the above selections.

The Viewpoint ROTATE option

This option requires two angles to be entered by the user:
a) the angle in XY plane from the X axis – the **view** direction
b) the angle from XY plane – the **inclination (tilt)**

1 Open your MODR2004\3DWFM drawing and:
   a) erase any dimensions and hatching
   b) leave all text items – they will act as a 'reference' as the model is viewed from different angles

2 Layer MODEL current, UCS BASE and SE Isometric viewpoint

3 Refer to Fig. 10.1A

4 At the command line enter **VPOINT <R>** and:
   prompt ***Switching to WCS***
   and   Current view direction: VIEWDIR=1.00,−1.00,1.00
   then   Specify a view point or [Rotate]<display compass and tripod>
   enter   **R <R>** – the rotate option
   prompt Enter angle in XY plane from X axis
   enter   **40 <R>**
   prompt Enter angle from XY plane
   enter   **0 <R>**
   prompt ***Returning to UCS***
   then   Regenerating drawing
   and   model displayed as Fig. 10.1(a1), i.e. looking towards the right-rear side from a horizontal 'stand-point' – the view direction

5 At the command line enter **VPOINT <R>** and:
   prompt Specify a view point or [Rotate]
   enter   **R <R>** – the rotate option
   prompt Enter angle in XY plane from X axis and enter: **90 <R>**
   prompt Enter angle from XY plane and enter: **0 <R>**
   and   model displayed as Fig. 10.1(a2)

6 Repeat the VPOINT command from the command line with the rotate option and enter the following angle values at the prompts:

prompt 1	*prompt 2*	*fig*
215	0	a3
330	0	a4

7 Restore the original SE Isometric viewpoint and refer to Fig. 10.1B

8 Use the VPOINT command with the rotate option, and enter the following angles at the prompts:

prompt 1	*prompt 2*	*fig*
0	45	b1
0	135	b2
0	270 (−90)	b3
0	−45 (315)	b4

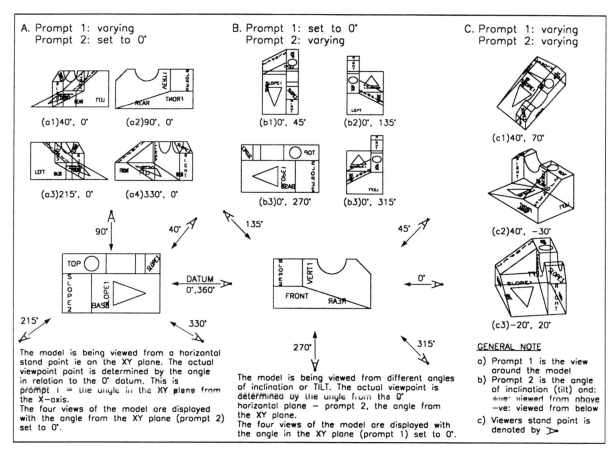

Figure 10.1   3D Views – the VPOINT Rotate option with 3DWFM.

9 Restore the SE Isometric viewpoint and refer to Fig. 10.1C. Activate the VPOINT Rotate command and enter the following angles:

prompt 1	prompt 2	fig
40	70	c1
40	−30	c2
−20	20	c3

10 Restore the original SE Isometric viewpoint

11 *Task*
Make some other saved UCS settings current, e.g. SLOPE1, VERT1, etc and repeat the Viewpoint Rotate command using some of the above angle entries. The model display should be unaffected by the UCS position. Think about the prompt ***Switching to the WCS***

12 *Explanation of option*

a) **Prompt 1: angle in XY plane from the X axis**
This is the viewer's stand-point **on the XY horizontal plane** looking towards the model, i.e. it is your view direction. If this angle is 0 degrees you are looking at the model from the right side. If the angle is 270 degrees you are looking onto the front of the model. The value of this angle can be between 0 and 360 degrees. It can also be positive or negative and remember that 270 degrees is the same as −90 degrees.

b) **Prompt 2: angle from the XY plane**
This is the viewer's 'head inclination' looking at the model, i.e. it is the **angle of tilt**. A 0 degrees value means that you are looking at the model horizontally and a 90 degrees value is looking vertically down. The angle of tilt can vary between 0 and 360 degrees and be positive or negative with:
positive tilt: looking down on the model
negative tilt: looking up at the model.

13 *Note*:
The reader must realise that the displays in Fig. 10.1 have been 'scaled' to fit the one sheet, and that your model displays will be larger than those illustrated.

VPOINT ROTATE using the presets dialogue box

1 3DWFM displayed at SE Isometric setting with UCS BASE?

2 Menu bar with **View-3D Views-Viewpoint Presets** and:
prompt Viewpoint Presets dialogue box – Fig. 10.2
with 1. viewing angle: absolute to WCS
2. angle from X axis: 315 – left-hand 'clock'
3. angle from XY plane: 35.3 – right-hand 'arc'

3 This dialogue box allows:
a) viewing angle to be absolute to WCS or relative to UCS
b) angles to be set by selecting clock/arc position
c) angles to be set by altering values at **From:** line
d) plan views to be set

4 Respond to the dialogue box with:
a) leave absolute to WCS
b) change the X axis angle from 315 to 150
c) change the XY plane angle from 35.3 to 10
d) pick OK
e) the model will be displayed at the entered viewpoint angles

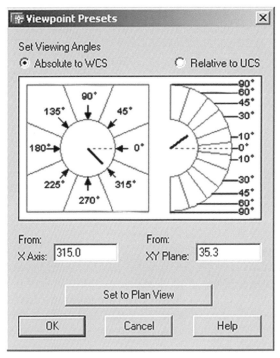

Figure 10.2 The Viewpoint Presets dialogue box.

5   Make UCS SLOPE1 current

6   Menu bar with View-3D Views-Viewpoint Presets and:
    a) make Relative to UCS active – black dot
    b) leave the two angle values as 150 and 10
    c) pick OK
    d) the model is displayed at the entered viewpoint angles but differs from the step 4
       display due to the UCS setting

7   *Task*
    a) Try some other entries from the Viewpoint Presets dialogue box using both selec-
       tion methods, i.e. the clock/arc and altering the angles
    b) Investigate the difference in the display with the Absolute to WCS and Relative to
       UCS selections
    c) Restore UCS BASE and the SE Isometric viewpoint

8   This completes the Viewpoint Rotate exercise. Do not save any changes to the 3DWFM
    model.

The Viewpoint COMPASS and TRIPOD option

This option allows the user to set 'infinite viewpoints'. Older users of AutoCAD will
remember this as the bulls-eye and target method. We will demonstrate the command
with a different model so:

1   Open the MODR2004\TEST3D model created during the viewport exercise and refer
    to Fig 10.3

2   Ensure UCS BASE is current and make the lower left viewport active, i.e the 3D
    Viewport

3   Menu bar with **View-Viewports-1 Viewport** to display a single viewport of the
    model at a 3D Viewpoint. This model 'fills the screen'.

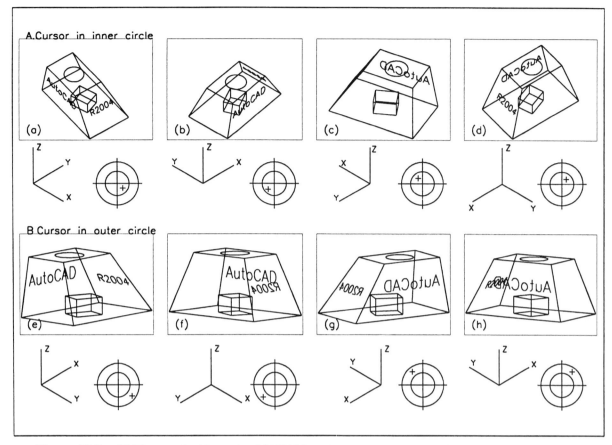

Figure 10.3 3D Views – the Viewpoint COMPASS and tripod option with TEST3D.

4   Menu bar with **View-3D Views-Viewpoint** and:
   prompt   1. model 'disappears'
             2. screen displays the XYZ tripod and the compass
             3. cursor replaced by a small cross (+)
             4. axes and cross (+) move as mouse is moved
   respond   move the cross (+) into the circle quadrant indicated in Fig. 10.3(a) and
             left-click
   and   model displayed at this viewpoint, and is viewed from above

5   At the command line enter **VPOINT <R>** and:
   prompt   ***Switching to WCS***
   then   Specify a viewpoint or [Rotate]<display compass and
             tripod>
   respond **press <RETURN>**
   prompt   tripod and compass displayed
   respond move the cross (+) into the circle quadrant indicated in Fig. 10.3(b) and
             left-click
   and   model displayed at this new viewpoint

6   Repeat the tripod viewpoint option (menu bar or command line) and position the
   cross (+) in the quadrants indicated in Fig. 10.3
   i.e. (c)–(d): within the inner circle
       (e)–(h): between the inner and outer circles

7 *Task*

When you are capable of using the compass and tripod, try the following:

a) position the + at different points on the two axes and observe the resultant displays

b) position the + at different points on the circle circumferences and note the displays

8 *Explanation of option*

a) The 'bulls-eye' is in reality a representation of a glass globe and the model is located at the centre of the globe. The XY plane is positioned at the equator. The north pole of the globe is at the circle centres and the two concentric circles represent the surface of the world, stretched out onto a flat plane with:

1. the circle centre: the north pole
2. the inner circle: the equator
3. the outer circle: the south pole

b) As the cross (+) is moved about the circles, the user is moving around the surfaces of the globe and:

Cross (+) position	View result
1. in inner circle	above equator, looking down on model
2. in outer circle	below equator, looking up at model
3. on inner circle	looking horizontally at model
4. below horizontal	viewing from the front
5. above horizontal	viewing from the rear

9 This completes the tripod option exercise. Do not save changes.

The Viewpoint VECTOR option

1 Open MODR2004\3DWFM with UCS BASE and SE Isometric viewpoint

2 Erase any dimensions and hatching, but leave the text items as they will act as a 'reference' as the model viewpoint is altered

3 Refer to Fig. 10.4

4 Menu bar with **View-3D Views-Viewpoint** and:

prompt	***Switching to WCS***
then	Current view direction
and	Specify a view point or [Rotate]
enter	**0,0,1 <R>**
prompt	***Returning to the UCS***
and	1. the model will be displayed at the entered viewpoint
	2. it is a top view – Fig. 10.4(a)
	3. it 'fills the screen'

6 At the command line enter **VPOINT <R>** and:

prompt	Specify a view point or [Rotate]
enter	**0,−1,0 <R>**
and	model displayed as Fig. 10.4(b) – a front view

7 Repeat the viewpoint vector option (menu bar or command line) and enter the following co-ordinates at the prompt line:

Vector entry	resultant view	fig (Fig. 10.4)
1,0,0	from right	c
0,1,0	from rear	d
−1,0,0	from left	e
0,0,−1	from below	f
1,1,1	3D from above	g
−1,−1,−1	3D from below	h

8 Restore the original SE Isometric viewpoint

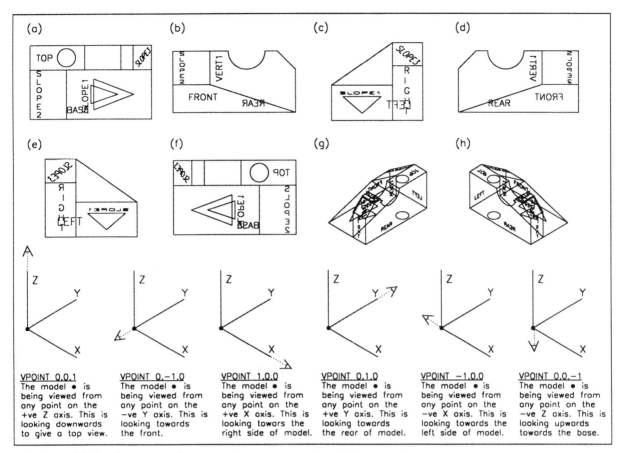

Figure 10.4 3D Views – the Viewpoint VECTOR option with 3DWFM.

9 *Task*

Try some vector entries for yourself then restore the original SE Isometric viewpoint

10 *Explanation of option*

a) The vector option allows the user to enter x,y,z co-ordinates. These are the co-ordinates of the viewers' 'stand-point' looking at the model which is considered to be at the origin. Thus if you enter 0,0,1 you are 'standing' at the point 0,0,1 looking towards the origin. As this point is on the positive Z axis you are looking down on the model, i.e. a top view.

b) The actual numerical value of the vector entered does not matter, i.e. 0,0,1; 0,0,12; 0,0,99.99; 0,0,3456 are all the same viewpoint entries. I prefer to use the number 1, hence 0,0,1; −1,0,0 etc.

c) Certain vector entries give the same display as rotate option and the following lists some of these similarities:

vector	rotate	view
0,0,1	0,90	top
0,−1,0	270,0	front
1,0,0	0,0	right
0,1,0	90,0	rear
−1,0,0	180,0	left
0,0,−1	0,−90	bottom
1,1,1	45,35	3D from above
−1,−1,−1	−135,−35	3D from below

11 This completes the vector option. Do not save any changes.

The isometric viewpoints

The isometric 3D Views allow the user to view a model from four 'preset' viewpoints, these being SW, SE, NE and NW. These four viewpoints are used extensively as they allow easy access to viewing a model in 3D.

1   Open model TEST3D to display the four viewport configuration saved from the previous chapter

2   Restore UCS BASE with layer MODEL current

3   Making the appropriate viewport active, menu bar with **Views-3D Views** and set the following viewpoints:

viewport   *viewpoint*
top left   SW Isometric
top right   SE Isometric
lower right   NE Isometric
lower left   NW Isometric

4   When the viewpoints have been entered, zoom-all in each viewport and the result should be Fig. 10.5. This exercise does not need to be saved.

5   *Notes*:

    a) The four preset isometric viewpoints only allow viewing from above. If a model is to be viewed from below, another option is required. My choice for this is VPOINT Rotate with a negative second angle value.

    b) The equivalent VPOINT Rotate angles for the four isometric presets are:

3D View	*angle in XY plane*	*angle from XY plane*
SW Isometric	225	35.3
SE Isometric	315	35.3
NE Isometric	45	35.3
NW Isometric	135	35.3

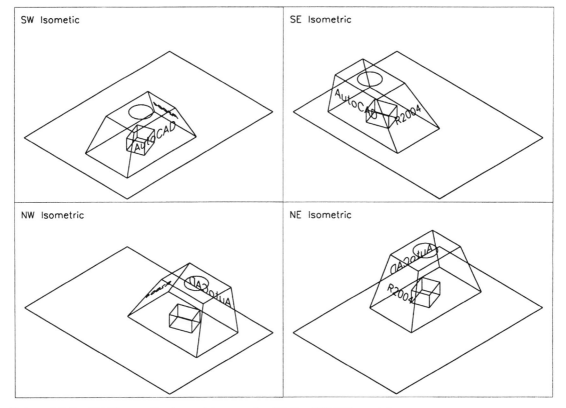

Figure 10.5   3D Views – the isometric presets with the TEST3D model.

The orthographic viewpoints

There are six 'preset' orthographic viewpoints these being Top, Bottom, Left, Right, Front and Back. The options are independent of the UCS position.

1 Open model 3DWFM and erase any dimensions and hatching

2 Restore UCS BASE with layer MODEL current. Refer to Fig. 10.6

3 Menu bar with **View-Viewports-2 Viewports** and:

   prompt   Enter a configuration option and enter: **H <R>**

4 With the top viewport active, menu bar with **View-Viewports-3 Viewports** and:

   prompt   Enter a configuration option and enter: **V <R>**

5 With the bottom viewport active, repeat step 4

6 Making each viewport active, menu bar with **View-3D Views** and set the following orthographic viewpoints:

viewport	*viewpoint*
top left	top
top middle	bottom
top right	left
lower right	right
lower middle	front
lower right	back

7 This exercise does not need to be saved.

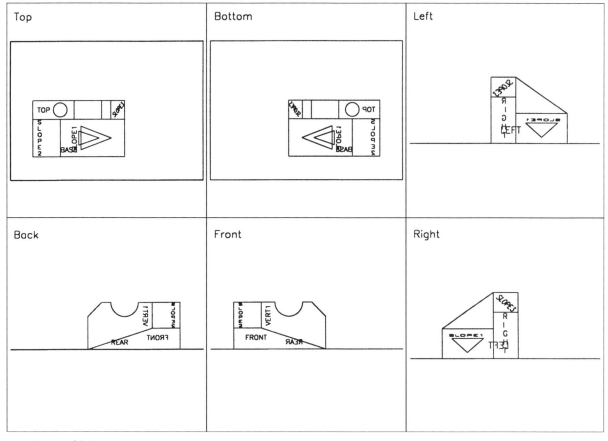

Figure 10.6   3D Views – the six orthographic presets with the 3DWFM model.

Viewpoint PLAN

This viewpoint selection was discussed during the chapter on the UCS and gives a view perpendicular to the current UCS position.

1 Open the four viewport configuration TEST3D and:
   a) create a single viewport configuration of the 3D View
   b) set any 3 viewport configuration
   c) make UCS FRONT current and refer to Fig. 10.7

2 Make any viewport active and menu bar with **View-3D Views-Plan View-Current UCS** and the model will be displayed as Fig. 10.7(a).

3 It is a plan view perpendicular to the UCS FRONT XY plane.

4 With another viewport active, menu bar with **View-3D Views-Plan View-World UCS** and the model will be displayed as Fig. 10.7(b)

5 Make the third viewport active and select the menu bar sequence **View-3D Views-Plan View-Named UCS** and:
   prompt Enter name of UCS and enter: **SLOPE <R>**

6 The model will be displayed as a view perpendicular to the XY plane of UCS SLOPE as Fig. 10.7(c)

7 This exercise does not need to be saved.

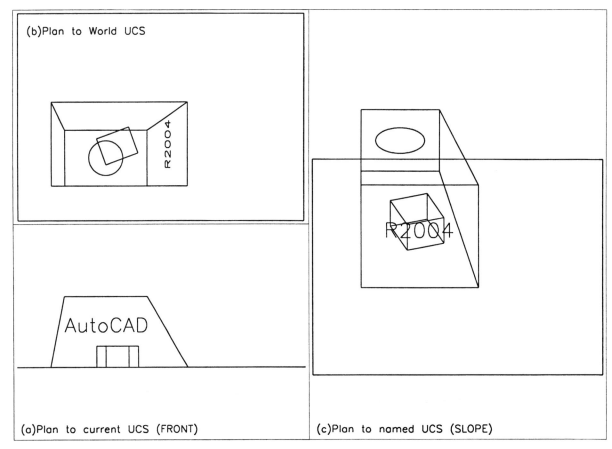

Figure 10.7 3D Views – the PLAN options with model TEST3D.

The VIEW command

Different views of a model can be saved within the current drawing, thus allowing the operator to create a series of 'pictures'. These could be of the model being constructed, of a completed model at differing viewpoints, etc. and these views (pictures) can be recalled at any time.

1 Open the 3DWFM model and erase any dimensions and hatching (or freeze the appropriate layers)

2 Model displayed at SE Isometric viewpoint with UCS BASE and layer MODEL current

3 At the command line enter **-VIEW <R>** and:
  prompt Enter an option [?/Orthographic/Delete/Restore/Save..
  enter **S <R>** – the save option
  prompt Enter view name to save
  enter **V1 <R>**

4 Menu bar with View-3D Views and set to NW Isometric

5 Menu bar with **View-Named Views** and:
  prompt View dialogue box – Named Views tab active
  respond **pick New**
  prompt New View dialogue box
  respond 1. enter View name: V2
       2. ensure Current display active, i.e. black dot
       3. Save UCS with view active (tick)
       4. UCS name: BASE – Fig. 10.8
       5. pick OK
  prompt View dialogue box with V2 added to list
  respond **pick OK**

6 Menu bar with View-3D Views and set to SW Isometric

7 Menu bar with **View-Named Views** and from the View dialogue box:
  a) pick New
  b) View name: V3 with current display, UCS BASE then OK
  c) View dialogue box as Fig. 10.9
  d) pick OK

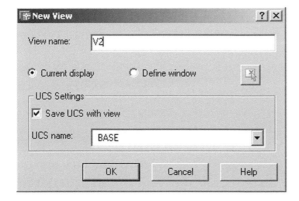

Figure 10.8 The New View dialogue box.

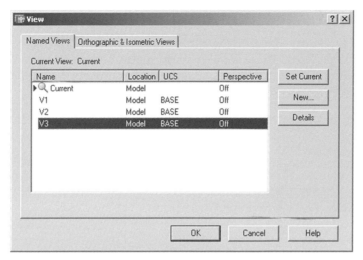

Figure 10.9 The View dialogue box (Named Views tab active).

8 At the command line enter **-VIEW <R>** and:
   prompt Enter an option
   enter **R <R>** – the restore option
   prompt Enter view name to restore
   enter **V1 <R>**
   and saved view V1 displayed

9 Menu bar with **View-Named Views** and:
   prompt View dialogue box
   respond 1. pick V3
             2. pick Set Current
             3. pick OK
   and screen displays the saved V3 view

10 *Task*
    a) create, save and display some other views
    b) investigate the Details option from the Named Views tab
    c) investigate the Orthographic & Isometric Views tab
    d) investigate altering the UCS with saved views

11 When complete, restore the SE Isometric viewpoint. Save if required but not as 3DWFM

12 *Note*:
    a) Do not confuse the VIEW command with the View option of the UCS command.
    They are two entirely different concepts
    b) The save option of the VIEW command is used when scenes have to be rendered.
    This will be demonstrated in the chapter on rendering solid models.

Viewports dialogue box

Viewpoints can be set using the various options of the 3D View command as well as from the Viewports dialogue box. To investigate the Viewports dialogue box further:

1 Open the 3DWFM model – no dimensions or hatching

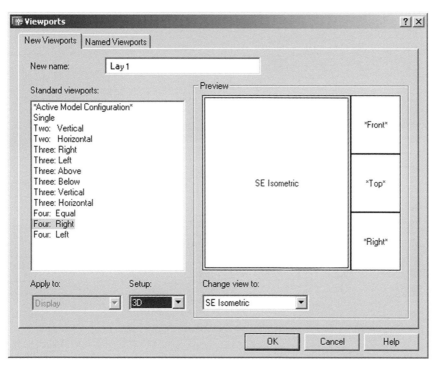

Figure 10.10 The Viewports dialogue box (New Viewports tab active).

2 UCS BASE and layer MODEL both current

3 Menu bar with **View-Viewports-Named Viewports** and:
prompt   Viewports dialogue box
respond   1. pick the New Viewports tab
          2. enter View name: **LAY1**
          3. pick Standard viewports: **Four: Right**
          4. Setup: 3D
          5. note Preview
          6. leave view names as given – Fig. 10.10
          7. pick OK

4 The screen will display a four viewport configuration with the appropriate views of the model as Fig. 10.11.

5 Repeat the menu bar sequence View-Viewports-Named Viewports and:
a) pick the New Viewports tab
b) enter View name as LAY2
c) pick a Four Left standard viewport
d) apply to display with 3D setup
e) pick OK
f) screen displays the same four views of the model as before, but in a different viewport configuration

6 Now restore the original single SE Isometric viewport layout

7 This completes this exercise which you can save if required.

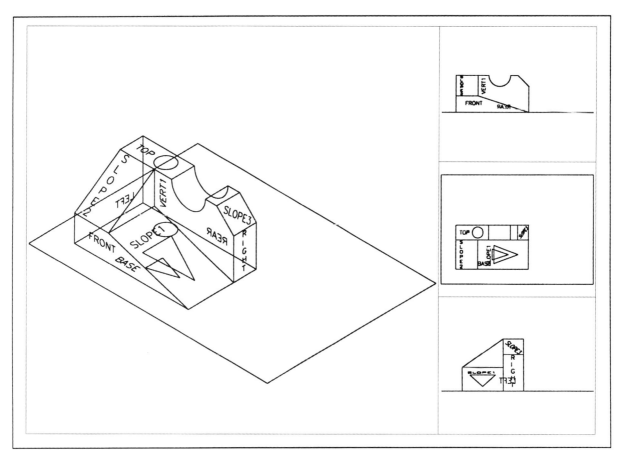

Figure 10.11   3DWFM using the Viewports dialogue box (New Viewports tab).

The View toolbar

All commands have so far been activated from the command line or the menu bar. The View toolbar has several icons which can be used to obtain several viewpoints of a model. The toolbar is displayed in Fig. 10.12. The View toolbar can be activated from the menu bar with **View-Toolbars** and 'ticking' the View toolbar. The Toolbars dialogue box can then be closed, and the View toolbar positioned to suit.

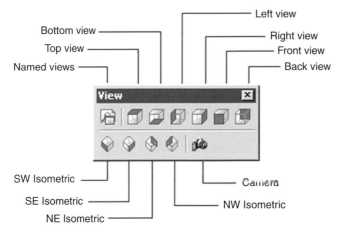

Figure 10.12   The View toolbar.

The user can now activate 3D commands by:

a) keyboard entry at the command line

b) using the menu bar selections

c) selecting an icon from the View toolbar

It is user preference as to which method is best suited to their own requirements.

Centring models in viewports

When 3D models are displayed in multiple viewport configurations, three 'problems' may initially occur:

a) the model may 'fill the viewport'

b) the model may be displayed at different sizes in each viewports

c) the model views may not 'line up' between viewports

These 'problems' are easily overcome by zooming each viewport by a scale factor or about a specified centre point determined by the user, who then decides on the 'scale effect' in the viewports. We will demonstrate the concept with three previously created models.

Example 1 – centring by scale

1   Open MODR2004\3DWFM of the wire-frame model and:

    a) erase/freeze any dimensions and text

    b) leave the hatching displayed

    c) erase the black border

    d) zoom-all and the model 'fills the screen'

    e) ensure UCS Icon is On and at Origin with menu bar View-Display-UCS Icon (these should be active)

    f) ensure UCS BASE is current

    g) command line with **UCSVP <R>** and:

        prompt   Enter new value for UCSVP

        respond enter: **0 <R>**

2   Menu bar with **View-Viewports-4 Viewports** and the model is displayed in 3D in each viewport

3   Menu bar with **View-3D Views** and set the following in the named active viewports:

viewport	*viewpoint*
top left	Right
top right	Front
lower right	Top
lower left	SE Isometric

4   The model will be displayed at the viewpoints entered and will be of differing sizes in each viewport. The model now needs to be centred about a specific point.

5   With the top left viewport active, menu bar with **View-Zoom-Scale** and:

    prompt   Enter a scale factor

    respond enter: **1.75 <R>**

6   Repeat the zoom-scale selection in the other three viewports, and enter a scale factor of 1.75 in the top right and lower right viewports, but 1 in the lower left (3D) viewport.

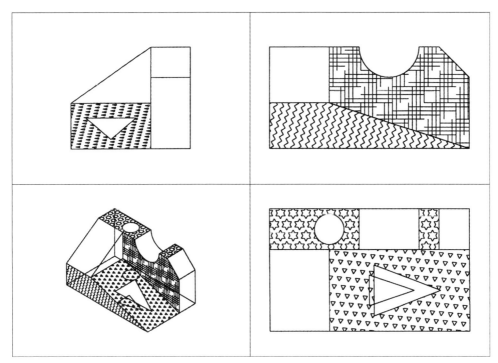

Figure 10.13   Centring viewport Example 1 – 3DWFM by scale factor.

7   When the zoom-scale command has been completed, the model will be 'neatly centred' in all viewports as Fig. 10.13

8   Save as MODR2004\MV3DWFM

Example 2 – centring about a user specified point

1   Open drawing TEST3D to display the four viewport configuration of the created model with text on two 'planes'

2   Erase the black border and zoom-all in each viewport and the model will be displayed at different 'sizes' in the viewports

3   Ensure UCS BASE is current

4   *a*) ensure UCS Icon is On and at Origin
    b) in each viewport use the command line UCSVP command and set the variable to 0

5   The model has a basic overall cuboid sizes of 200×120×100 and its 'centre point' *relative to UCS BASE* is at 100,60,50.

6   With the top left viewport active, menu bar with **View-Zoom-Centre** and:
    prompt   Specify centre point
    enter   **100,60,50 <R>**
    prompt   Enter magnification or height <some value>
    enter   **175 <R>**

7   Repeat the zoom-centre command in the other three viewports and enter the centre point as 100,60,50 and the magnification as 175 but 225 in the 3D Viewport

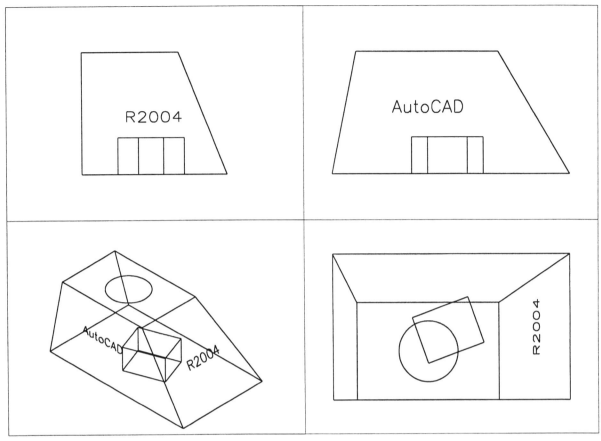

Figure 10.14 Centre viewport Example 2 – TEST3D by centre/magnification.

8 The model will be centred in each viewport as Fig. 10.14

9 Save this display as MODR2004\MVTEST3D.

Example 3 – centring with zoom-extents

1 Open the second saved 2½D model from Chapter 2. This model should *be displayed in 3D at a SE Isometric viewpoint*

2 Set a 4 viewport configuration with the menu bar sequence **View-Viewports-4 Viewport**

3 Menu bar with **View-3D Views** and set the following in the named active viewports:

viewport	*viewpoint*
top left	Right
top right	Front
lower right	Top
lower left	SE Isometric

4 Menu bar with **View-Zoom-Extents** and the model will be displayed at different sizes in each viewport

5 At the command line enter ZOOM then 1 in all viewports. If the model is not displayed 'to your satisfaction', try the ZOOM command with other values, e.g. 0.75, 1.25, 1.5 etc.

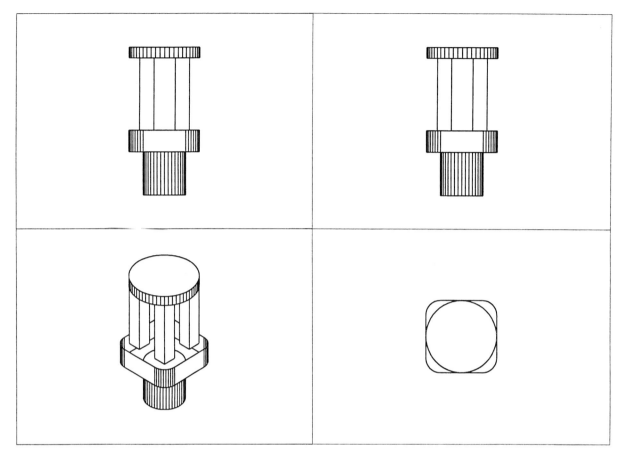

Figure 10.15   Centre viewport Example 3 – Extruded model using Zoom-Extents.

6   When complete the screen layout should be as Fig. 10.15, which is displayed with the hide effect. The layout can be saved if required.

Summary

1   The viewpoint command allows models to be viewed from different 'stand-points'

2   The command has several selection options including:
   a)   four preset isometric views – SW, SE, NE and NW
   b)   six orthographic preset views – top bottom, left, right, front and back
   c)   three plan views – current UCS, world UCS, named UCS
   d)   VPOINT with three options – rotate, compass and tripod, vector

3   The VPOINT Rotate option requires two angles:
   a)   the angle 'around' the model – the direction
   b)   the angle of inclination – the tilt

4   The VPOINT Rotate option can be set from a dialogue box

5   The VPOINT compass/tripod option allows unlimited viewpoints

6   The VPOINT vector option requires an x,y,z co-ordinate entry

7   Viewpoints are generally set **absolute to the WCS** and the relative to the UCS option is **not recommended**

8 All wire-frame models exhibit **ambiguity** when the viewpoint command is used, i.e. viewed from above or from below?

9 The VIEW command allows different views of a model to be saved in the current drawing for future recall. This is useful when the model is being displayed at various viewpoints.

10 Models can be centred in multiple tiled viewports using:
   a) Zoom-scale, the user entering the scale factor
   b) Zoom-centre, the user entering the centre point and the magnification. This centre point is **dependent on the UCS position**.
   c) Zoom-Extents then zoom with a value entered by the user

11 I personally prefer to either use:
   a) the Zoom-Extents method
   b) Zoom-centre relative to UCS BASE which is usually 'set' at a convenient base vertex.

12 The magnification value entered is a 'scale' effect and is relative to the given default, e.g. if the default is <180> then:
   a) a value less than 180 will increase the model size
   b) a value greater than 180 will decrease the model size.

Assignment

A single activity has been included at this stage, which involves creating multiple tiled viewports, setting viewpoints and centring the model. The model has already been created and saved (hopefully) during the hatching activities.

Activity 8: The hatched pyramid of MACFARAMUS.

1 Open the hatched pyramid model from *Activity 7* and erase the border

2 Using the New Viewports dialogue box, create a four viewport (left) configuration to display a SE Isometric view as well as front, top and right views

3 Centre the model in the viewports (Zoom-Extents method is easy)

4 This activity should take no more than 5 minutes?

5 When complete, save as MODR2004\MVPYR

Chapter 11

Model space, paper space and untiled viewports

AutoCAD 2004 has multi-view capabilities which allow the user to layout, organise and plot multiple views of any 3D model. The multiple viewport concept has already been used with our wire-frame models, these viewports being TILED, i.e. fixed.

In this chapter we will investigate how to create UNTILED or FLOATING viewports which are used in the same way that the tiled viewports were used. The creation of untiled viewports requires an understanding of the two AutoCAD drawing environments – model space and paper space.

The basic model/paper space concepts are, referring to Fig. 11.1:

1   *Model space*
    This is the drawing environment that exists in any viewport and is the default. All models that have so far been created have been completed in model space. Model space is used for all draughting and design work and for setting 3D viewpoints. Multiple viewports are possible in model space but are TILED, i.e. they cannot be moved or altered in size – Fig. 11.1(a). While model space multi-views are useful, they have one major disadvantage – only the active viewport can be plotted, i.e. model space multiple viewports cannot be plotted on one sheet of paper.

2   *Paper space*
    This is a drawing environment which is independent of model space. In paper space the user creates the drawing sheet, i.e. border, title box, etc. as well as arranging the multiple viewport layout. The viewports created in paper space are UNTILED, i.e. they can be positioned to suit, altered in size and additional viewports can be added to the layout – Fig. 11.1(b). In paper space the 3D viewpoint command is not valid although objects (particularly text) can be added to the sheet layout. The real advantage of working with paper space multiple viewports is that any viewport configuration can be plotted on the one sheet of paper.

3   *Tilemode*
    The system variable which controls the 'type' of viewport to be created is **TILEMODE** and:
    a) TILEMODE 1: model space (FIXED) viewports and paper space are not available
    b) TILEMODE 0: paper space (FLOATING) viewports and model space are both available

    Tiled (model space) viewports are always displayed as edge-to-edge and fill the screen like a tiled wall. Untiled (paper space) viewports can be positioned anywhere within the screen area with spaces between them if required. They can also be copied, moved, stretched, etc.

4   *Icons*
    When working in model space the normal WCS/UCS icon will be displayed in all viewports, orientated to the viewport viewpoint as Fig. 11.1(c). In paper space, the paper space icon – Fig. 11.1(d) is displayed.

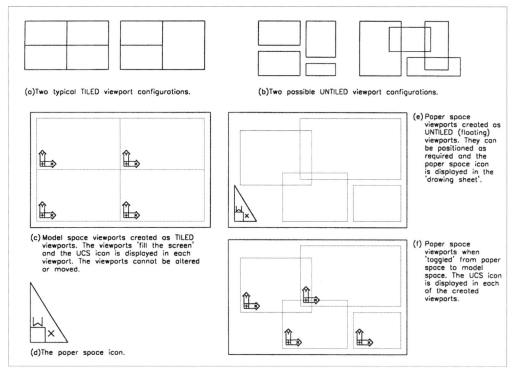

Figure 11.1   Model and paper space concepts.

When viewports are created in paper space, the paper space icon is displayed in the lower left corner of the screen as Fig. 11.1(e), but when model space 'is entered' the UCS icon is again displayed in all created viewports as Fig. 11.1(f).

5   *Toggling between model and paper space*
All AutoCAD users will be familiar with the toggle concept, e.g. toggling the grid on/off with the F7 key or from the Status bar with a left-click on the word GRID. It is possible to toggle between model space and paper space but only if the TILEMODE system variable has been set to 0. The toggle effect can be activated by:
1. *Command line*
   a) if in paper space, toggle to model space with **MS <R>**
   b) if in model space, toggle to paper space with **PS <R>**
2. *Status bar*
   a) left-click on PAPER to toggle to model space
   b) left-click on MODEL to toggle to paper space

6   *UCSVP*
UCSVP is a system variable which determines the orientation of the UCS in multiple viewports and:
a) UCSVP 0 : the UCS icon reflects the UCS setting of the current viewport
b) UCSVP 1 : the UCS icon is independent of the UCS setting of the current viewport

This means that if the UCSVP is set to 0 and the UCS is altered, the icon will reflect this altered position. If UCSVP is set to 1, and alteration of the UCS will not be reflected by the icon. For all our multiple viewport work, **UCSVP will be set to 0**, as I believe the icon should reflect the UCS setting.

7   *Layout tabs*
AutoCAD has a layout tab line which allows the user to create several layouts of a model in the current drawing. The layout tab line has a Model tab display and can have several layout tabs. We will use the Model and Layout tabs with our multiple viewports.

Model/paper space example – untiled viewports

This example will use a model created in model space to demonstrate the paper space multiple viewport concept. The example is quite long but if you are unsure of paper space, persevere with it – it is important that you understand how to create and use paper space viewports.

Note:

a) Although you may be familiar with the model/paper space concept it is advisable that you complete the exercise

b) AutoCAD 2004 assumes that the user has access to a printer or plotter, this being connected to the computer. As all printer/plotters differ in their configurations, paper size, etc. I have **assumed** that no printer is configured. This does not exclude the user from obtaining a hard copy of completed drawings, it is simply a matter of convenience for me when detailing the various steps in the exercises.

c) There are different methods of creating paper space untiled viewports from a model space drawing. This example demonstrates one of these methods.

1   Open drawing MODR2004\3DWFM and:
   a) erase the black border, hatching and any dimensions
   b) leave the text items – they will 'act as a reference'
   c) layer MODEL and UCS BASE current
   d) zoom-all and model 'fills the screen'
   e) refer to Fig. 11.2

2   Make a new layer called VP, linetype continuous, colour 12

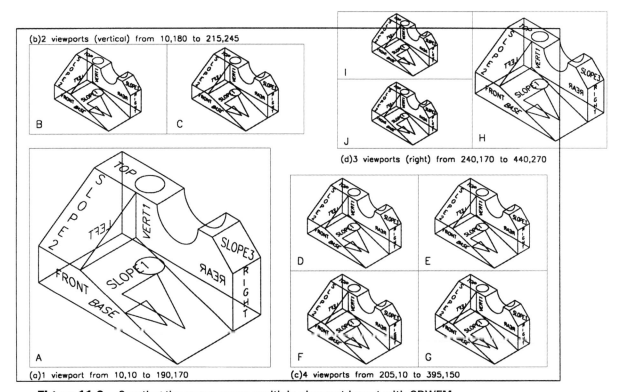

Figure 11.2   Creating the paper space multiple viewport layout with 3DWFM.

3 At the command line enter **PS <R>** and prompt line will display:

** Command not allowed in Model tab **

4 Menu bar with **Tools-Wizards-Create Layout** and:

prompt   Create Layout - Begin dialogue box
respond enter layout name: **MYFIRST** then pick Next
prompt   Create Layout - Printer dialogue box
respond pick **None** then pick Next
prompt   Create Layout - Paper Size dialogue box
respond 1. scroll and pick **ISO A3 (420.00 × 297.00 MM)**
           2. Drawing units: millimetres
           3. pick Next
prompt   Create Layout - Orientation dialogue box
respond ensure **Landscape** active then pick Next
prompt   Create Layout - Title Block dialogue box
respond pick **None** then pick Next
prompt   Create Layout - Define Viewports dialogue box
respond pick **None** then pick Next
prompt   Create Layout - Finish dialogue box
respond pick **Finish**
and   1. screen returned in paper space – note the icon
           2. the white area is the A3 paper
           3. the dotted area is the permitted drawing area within the white A3 paper – for plot purposes
           4. a new layout name MYFIRST is added to the tab bar and is active, i.e. highlighted

5 Make layer 0 current and menu bar with **Draw-Rectangle** and:

prompt   Specify first corner point
enter   **0,0 <R>**
prompt   Specify other corner point
enter   **405,257 <R>**

6 This rectangle is our drawing area outline

7 Make layer VP current

8 Menu bar with **View-Viewports-1 Viewport** and:

prompt   Specify corner of viewport and enter: **10,10 <R>**
prompt   Specify opposite corner and enter: **190,170 <R>**
and   a viewport (A) is created with the model 3DWFM displayed as Fig. 11.2(a)

9 Menu bar with **View-Viewports-2 Viewports** and:

prompt   Enter viewport arrangement and enter: **V <R>**
prompt   Specify first corner and enter: **10,180 <R>**
prompt   Specify opposite corner and enter: **215,245 <R>**
and   two additional viewports (B and C) are created, each displaying the model in 3D – Fig. 11.2(b)

10 Menu bar again with **View-Viewports-4 Viewports** and:

prompt   Specify first corner and enter: **205,10 <R>**
prompt   Specify opposite corner and enter: **395,150 <R>**
and   four new viewports (D,E,F,G) are created with the model displayed in each – Fig. 11.2(c)

11 Final menu bar selection with **View-Viewports-3 Viewports** (right option) and:
   prompt Specify first corner and enter: **240,170 <R>**
   prompt Specify opposite corner and enter: **440,270 <R>**
   and   three viewports (H,I,J) are created each with the model displayed as
             Fig. 11.2(d). These viewports extend 'outwith' the drawing paper

12 *What has been achieved?*
   a) ten viewports have been created
   b) the viewports have been created in paper space
   c) each viewport displays the original 3DWFM model

13 At the command line enter MS <R> and:
   a) toggled to model space
   b) the last viewport (top right) is active?
   c) each viewport displays the UCS icon at BASE?
   d) any viewport can be made active as before, i.e. move to required viewport and
       left-click

14 Investigate the paper/model space toggle:
   a) from the command line with PS and MS
   b) status bar with left-click on MODEL/PAPER
   c) decide on which toggle method you prefer

15 Investigate the model/layout tab line:
   a) pick the Model tab and the original 3DWFM model is displayed
   b) pick the MYFIRST tab and our multiple viewport layout is displayed

16 At this stage the layout should be similar to Fig. 11.2.

Setting the viewpoints

1 Enter model space and set the following viewpoints in the named viewports with
  View-3D Views from the menu bar:

viewport	viewpoint	viewport	viewpoint
A	SW Isometric	F	SE Isometric
B	NE Isometric	G	Top
C	NW Isometric	H	Bottom
D	Right	I	Left
E	Front	J	Back

2 The model is now displayed at a different viewpoint in each viewport. Using
  the Zoom-Scale command, enter the following scale factors in the appropriate
  viewport:
  a) scale factor of 1.75 in orthogonal view viewports
  b) scale factor of 1.25 in 3D view viewports

3 The UCS icon may not be displayed as expected so:
  a) in each viewport, enter **UCSVP <R>** then **0 <R>**
  b) make the largest viewport (A) current
  c) restore UCS BASE and the icon should be displayed as expected in all
      viewports.

Adding text

1 In model space, make layer TEXT current and viewport (A) active

2 Using the menu bar sequence, **Draw-Text-Single Line Text**, add an item of text
using the following:
a) start point: 0, −25
b) height: 12.5 and rotation: 0
c) text item: AutoCAD 2004 (MS)

3 This item of text will be displayed in the ten viewports at different orientations due to
the viewpoints. In some viewports (D,E,I,J) the text is viewed 'end-on'

4 Enter paper space with PS <R>

5 Menu bar with **Draw-Text-Single Line Text** and add an item of text using:
a) start point: centred on 200,150
b) height: 20 and rotation: 0
c) text: AutoCAD 2004<R>
text: in <R>
text: PAPER SPACE <R><R> – two returns

6 At this stage your screen layout should resemble Fig. 11.3.

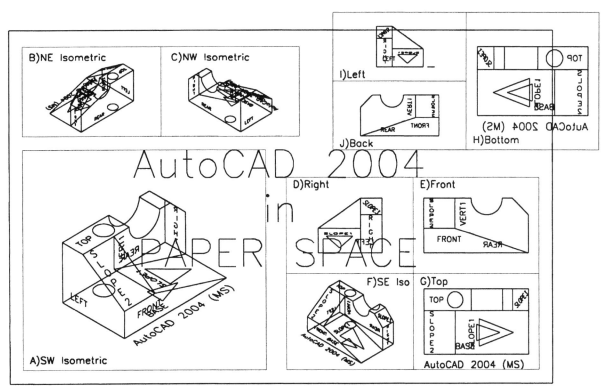

Figure 11.3 Working with the created paper space viewports.

Modifying the layout

1   In paper space try and erase the model – you cannot

2   In model space try and erase the paper space text – not possible

3   In paper space, activate the ZOOM command and window viewport A. The viewport
    will be enlarged and by entering model space, it is easier to work on the model

4   In paper space, zoom previous to restore the original layout

5   In paper space, select the SCALE icon from the Modify toolbar and:
    prompt   Select objects
    respond **completely window viewports H,I,J then right-click**
    prompt   Specify base point
    respond **Intersection of lower left corner of viewport J**
    prompt   Specify scale factor
    enter   **0.75 <R>**

6   In model space:
    a) zoom-all in viewports H, I and J
    b) then zoom-scale with the same 1.75 scale factor as before

7   Enter paper space and freeze the layer VP

8   The ten views of the model will be displayed with text but without the viewport
    borders – Fig. 11.4

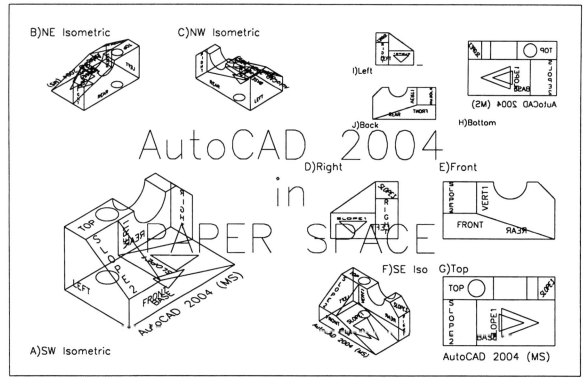

Figure 11.4   Completed paper space exercise with 3DWFM.

9 If you have access to a printer/plotter:
   a) print from any viewport in model space
   b) print from paper space

10 This completes this exercise – save if required.

Comparison between model space and paper space

The following table gives a brief comparison between the model and paper space drawing environments:

Model space	*Paper space*
used to create the model	used to create the paper layout
model can be modified	model cannot be modified
tiled (fixed) viewports	untiled (floating) viewports
tilemode: 1	tilemode: 0
viewports restricted in size	viewports to any size
viewports 'fill screen'	viewports positioned to suit
viewports cannot be altered	viewports can be moved, copied, etc.
cannot add viewports	additional viewports can be created
plot only active viewport	all viewports can be plotted
3D views active	cannot use 3D viewpoint
WCS or UCS icon	paper space icon
zoom in active viewport	zoom complete viewports

Finally

One of the major benefits of paper space is the ability to zoom a complete viewport. This allows the zoomed viewport to be enlarged, thus allowing the user to 'see more clearly' the model being created/worked on when in model space. This is a concept with which the user should become familiar and the procedure is:

1 Enter paper space with PS <R>

2 Zoom-window a specific viewport, e.g. the 3D viewport

3 Return to model space with MS <R>

4 Complete (or modify) the model

5 Enter paper space with PS <R>

6 Zoom previous (or All) to restore the complete layout

7 Return to model space with MS <R>

This zoom a viewport concept in paper space is a very useful concept and will be used in all future exercises.

Chapter 12

Updating the A3 standard sheet

Now that the model/paper space concept has been discussed, a new standard sheet will be created which will allow all future models (surface and solid) to be displayed in multiple viewports. This standard sheet will be created with three layout tab displays, and the original 3DSTDA3 sheet will be modified as it:
a) already has layers, e.g. MODEL, OBJECTS, TEXT, etc.
b) has a created dimension style – 3DSTD
c) other variables set.

Getting ready

1 Close any existing drawings and open your 3DSTDA3 standard sheet, either template or drawing (it does not matter) to display:
a) the black border at a SE Isometric viewpoint
b) the WCS icon at the left vertex of the border

2 Erase the black border – we will not use it again

3 Menu bar with **View-Display-UCS Icon** and:
a) On and Origin active (tick)
b) Properties and decide on a 2D or 3D icon display

4 Menu bar with **Format-Layer** and create two new layers:
name	*colour*	*linetype*
VP	number 14	continuous
SHEET	number 212	continuous

5 At the command line enter **UCSVP <R>** and:
prompt   Enter new value for UCSVP
enter   **0 <R>**

6 Note the layout tab line – similar to Fig. 12.1(a)

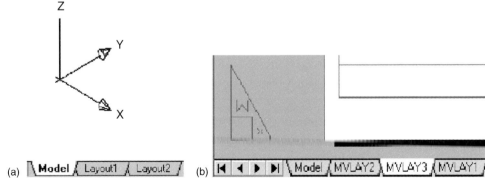

(a) (b)

Figure 12.1   The layout tab displays. (a) Original layout tab display with a 3D icon. (b) The layout tab display after creating three new viewport configurations with paper space active.

The first layout tab display

1 Menu bar with **Tools-Wizards-Create Layout** and with the Create Layout dialogue box set the following:
   a) Begin: enter **MVLAY1** then Next
   b) Printer: None then Next
   c) Paper Size: Millimetres, ISO A3 (420.00 × 297.00 MM) then Next
   d) Orientation: Landscape then Next
   e) Title Block: None then Next
   f) Define viewports: None then Next
   g) Finish: pick Finish

2 A white paper area will be returned with a dotted drawing area within, this being in paper space

3 Make layer SHEET current

4 Menu bar with **Draw-Rectangle** and:
   prompt   Specify first corner point and enter: **0,0 <R>**
   prompt   Specify other corner point and enter: **405,257 <R>**

5 Zoom-all and PAN to suit

6 Draw a line from: 0,15 to: @405,0

7 The area at the bottom of the 'paper' is for you to 'customise' as required. Use layer SHEET for this.

8 Make layer VP current

9 Menu bar with **View-Viewports-New Viewports** and:
   prompt   New Viewports dialogue box
   respond 1. pick Four: Equal
              2. Viewport spacing: 0
              3. Setup: 3D
              4. Change views to: leave unaltered
              5. pick OK

10 The screen will display four paper space viewports within the sheet border

11 To set and save some UCS positions, enter model space with the command line entry **MS <R>**

12 Making each viewport active, enter **UCSVP <R>** and:
   prompt   Enter new value for UCSVP
   enter   **0 <R>**

13 Menu bar with **Tools-New UCS-Origin** and:
   prompt   Specify new origin point
   enter   **10,10,0 <R>**

14 At command line enter **UCS <R>** then **S <R>** and:
   prompt   Enter name to save current UCS
   enter   **BASE <R>**

15 Menu bar with **Tools-New UCS-X** and:
   prompt   Specify rotation angle about X axis
   enter   **90 <R>**

16 Command line with **UCS <R>** then **S <R>** and enter: **FRONT <R>**

17 Menu bar with **Tools-New UCS-Y** and:
prompt Specify rotation angle about Y axis
enter **90 <R>**

18 Command line with **UCS <R>** then **S <R>** and enter: **RIGHT <R>**

19 We have now set and saved three UCS positions – BASE, FRONT and RIGHT. These three UCS positions will assist with future model creation.

20 Restore UCS BASE and make the lower left viewport active.

A second layout tab display

1 Still with the MVLAY1 tab and the lower left viewport active with UCS BASE and layer VP current

2 Left-click on the Layout1 tab and:
prompt Page Setup – Layout1 tab dialogue box
respond 1. Plot Device tab active and set:
   a) Plotter configuration: None
   2. Layout Settings tab active and set:
   a) Paper size: ISO A3 (420.00 × 297.00 MM)
   b) Drawing orientation: Landscape
   c) Plot area: Layout
   d) Plot scale: 1:1
   e) pick OK
and Paper space entered with:
   a) white drawing paper
   b) dashed line plotting area
   c) coloured rectangle – a new viewport
respond erase the viewport

3 Make layer SHEET current

4 Menu bar with **Draw-Rectangle** and:
prompt Specify first corner point and enter: **0,0 <R>**
prompt Specify other corner point and enter: **405,257 <R>**

5 Zoom-all and PAN to suit

6 Draw a line from: 0,15 to: @405,0 and customise the area below this line using layer SHEET

7 Make layer VP current and menu bar with **View-Viewports-New Viewports** and set:
a) Configuration: Two vertical
b) Spacing: 0
c) Setup: 3D
d) Change view to: leave at present
e) pick OK
f) At first corner prompt: enter 10,20
g) Opposite corner prompt: enter 395,247

8 Enter model space and:
a) set UCSVP to 0 in each viewport
b) restore UCS BASE

9 Make the right viewport active and with **VPOINT <R>** at the command line:
a) enter R <R> for the rotate option
b) enter 210 as the angle in the XY plane
c) enter 30 as the angle from the XY plane

10 With the left viewport active, **VPOINT <R>** again and:
   a) enter R <R> for the rotate option
   b) enter 210 as the angle in the XY plane
   c) enter −30 as the angle from the XY plane

11 We have set two viewpoints in this layout tab, one from above (the right viewport) and one from below (the left viewport)

12 Right-click on Layout1 tab name and:
   prompt   Shortcut menu
   respond **pick Rename**
   prompt   Rename Layout dialogue box
   respond 1. alter name to **MVLAY2**
              2. pick OK

13 Now have created two layout settings, MVLAY1 and MVLAY2 and the Model tab still available.

A third layout tab display

Although we already have two layout tabs, we will create a third layout configuration which is slightly different from the two 'traditional' layouts MVLAY1 and MVLAY2. The sequence is long and uses co-ordinates for most of the entries. I would ask the user to persevere with this exercise, as the end result is worth the time and effort spent.

1 Still with the MVLAY2 tab, the right viewport active UCS BASE and layer VP current

2 Left-click on the Layout2 tab and:
   prompt   Page Setup – Layout2 tab dialogue box
   respond 1. Plot Device tab active and set:
                 a) Plotter configuration: None
              2. Layout Settings tab active and set:
                 a) Paper size: ISO A3 (420.00 × 297.00 MM)
                 b) Drawing orientation: Landscape
                 c) Plot area: Layout
                 d) Plot scale: 1:1
                 e) pick OK
   and   Paper space entered with:
              a) white drawing paper
              b) dashed line plotting area
              c) coloured rectangle – a new viewport
   respond erase the viewport

3 Make layer SHEET current

4 Menu bar with **Draw-Rectangle** and:
   prompt   Specify first corner point and enter: **0,0 <R>**
   prompt   Specify other corner point and enter: **405,257 <R>**

5 Zoom-all and PAN to suit

6 Draw a line from: 0,15 to: @405,0 and customise the area below this line using layer SHEET

7 Make layer VP current

8 Menu bar with **View-Viewports-Polygonal Viewport** and:
   prompt   Specify start point and enter: **5,140 <R>**
   prompt   Specify next point and enter: **140,140 <R>**
   prompt   Specify next point and enter: **170,195 <R>**
   prompt   Specify next point and enter: **170,250 <R>**
   prompt   Specify next point and enter: **5,250 <R>**
   prompt   Specify next point and enter: **C <R>** – the close option

9 A five sided viewport will be displayed at the top left corner of the drawing area

10 Menu bar with **Modify-Mirror** and:
   prompt   Select objects
   respond **pick any point on viewport border** then right-click
   prompt   First point of mirror line and enter: **202.5,15 <R>**
   prompt   Second point of mirror line and enter: **202.5,250 <R>**
   prompt   Delete source objects and enter: **N <R>**

11 Repeat the menu bar with **Modify-Mirror** selection and:
   prompt   Select objects
   respond **pick any point on the two viewports** then right-click
   prompt   First point of mirror line and enter: **0,136 <R>**
   prompt   Second point of mirror line and enter: **405,136 <R>**
   prompt   Delete source objects and enter: **N <R>**

12 We have now created four polygonal viewports

13 Menu bar with **View-Viewports-Polygonal Viewport** and:
   prompt   Specify start point and enter: **152.5,136 <R>**
   prompt   Specify next point or [Arc/Length/Undo]
   enter   **A <R>** the arc option
   prompt   Enter arc boundary option
   enter   **CE <R>** – the centre point option
   prompt   Specify center point of arc and enter: **202.5,136 <R>**
   prompt   Specify endpoint of arc and enter: **252.5,136 <R>**
   prompt   Specify endpoint of arc and enter: **152.5,136 <R>**
   prompt   Specify endpoint of arc and **<RETURN>** to end command

14 We have now added a circular viewport to our Layout3 display

15 Enter model space and make any of the new viewports active

16 Menu bar with **View-Display-UCS Icon** and ensure that On and Origin are both active. I had to complete this sequence in all the five viewports.

17 In all viewports set UCSVP to 0

18 Make layer MODEL current and restore UCS BASE

19 Set five viewpoints to your own specification. I set the viewpoints to SE Isometric, Left, Right, Front and Top, but it is your choice

20 Right-click on Layout2 tab and:
   a) pick Rename from shortcut menu
   b) enter **MVLAY3** as the layout name
   c) pick OK

21 At this stage the layout tab should display Model, MVLAY1, MVLAY2 and MVLAY3 similar to Fig. 12.1(b)

22 If you want the layout tabs to 'be in order' then right-click an MVLAY name and use the Move/Copy option of the shortcut menu to position as required.

Saving the layouts as a new standard sheet

1 Make any layout tab MVLAY1 (for example) current

2 Enter model space with the lower left viewport active

3 Layer MODEL and UCS BASE current

4 Menu bar with **File-Save As** and:
   prompt   Save Drawing As dialogue box
   respond 1. scroll at Files of Type
              2. pick **AutoCAD Drawing Template File (*.dwt)**
   prompt   list of existing template files in the AutoCAD template folder
   respond 1. enter File name as: **MV3DSTD**
              2. pick Save
   prompt   Template Description dialogue box
   respond 1. enter: **My multi-view 3D prototype layout drawing created on XX/YY/ZZ**
              2. Measurement: Metric
              3. pick OK

5 Repeat step 4, but save as a template to your named folder with the same name, i.e. MV3DSTD

6 Menu bar with **File-Save As** and:
   prompt   Save Drawing As dialogue box
   respond 1. scroll and pick **AutoCAD 2004 Drawing (*.dwg)**
              2. scroll and pick your named folder (MODR2004)
              3. enter file name: **MV3DSTD**
              4. pick Save

7 This saved template/drawing file will be used extensively when starting new model exercises.

Checking the new MV3DSTD layout

Now that the MV3DSTD template file has been created we will add some 3D objects to 'check' the layout. Try and reason out the co-ordinate entries.

1 Close any existing drawings

2 Menu bar with **File-New** and:
   prompt   Create New Drawing dialogue box
   respond 1. pick Use a Template
              2. scroll and pick **MV3DSTD.dwt** – Fig. 12.2
              3. pick OK

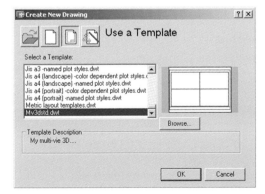

Figure 12.2   The Create New Drawing (Use a Template) dialogue box.

3   Your multiple viewport drawing should be displayed as saved, i.e. with a layout tab active, e.g. MVLAY1

4   Display the Object Snap and Surfaces toolbars

5   Menu bar with **Draw-Surfaces-3D Surfaces** and:
   prompt   3D Objects dialogue box
   respond **pick Box3d then OK**
   prompt   Specify corner of box and enter: **0,0,0 <R>**
   prompt   Specify length of box and enter: **200 <R>**
   prompt   Specify width of box and enter: **100 <R>**
   prompt   Specify height and enter: **80 <R>**
   prompt   Specify rotation angle about Z axis and enter: **0 <R>**

6   Menu bar with **Draw-Surfaces-3D Surfaces** and:
   prompt   3D Objects dialogue box
   respond **pick Wedge then OK**
   prompt   Specify corner point of wedge and enter: **0,0,0 <R>**
   prompt   Specify length of wedge and enter: **100 <R>**
   prompt   Specify width of wedge and enter: **100 <R>**
   prompt   Specify height of wedge and enter: **100 <R>**
   prompt   Specify rotation angle of wedge about Z axis and enter: **−90 <R>**

7   At the command line enter **CHANGE <R>** and:
   prompt   Select objects
   respond **pick the wedge then right-click**
   prompt   Specify change point or [Properties] and enter: **P <R>**
   prompt   Enter property to change and enter: **C <R>**
   prompt   Enter new color and enter: **BLUE <R><R>** – two returns

8   Select the CONE icon from the Surfaces toolbar and:
   prompt   Specify center point for base of cone and enter: **70,50,80 <R>**
   prompt   Specify radius for base of cone and enter: **50 <R>**
   prompt   Specify radius for top of cone and enter: **0 <R>**
   prompt   Specify height of cone and enter: **100 <R>**
   prompt   Enter number of segments for surface of cone and enter: **16 <R>**

9   Change the colour of the cone to green

10 Select the DISH icon from the Surface toolbar and:
   prompt   Specify center point of dish and enter: **150,50,0 <R>**
   prompt   Specify radius of dish and enter: **50 <R>**
   prompt   Enter number of longitudinal segments for surface of dish
   enter   **16 <R>**
   prompt   Enter number of latitudinal segments for surface of dish
   enter   **8 <R>**

11 Change the colour of the dish to magenta

12 Make each viewport current and:
   a) zoom to extents
   b) zoom to a scale of 1

13 With UCS BASE make layer TEXT current and menu bar with **Draw-Text-Single Line Text** and:
   prompt   Specify start point of text and enter: **130,80,80 <R>**
   prompt   Specify height and enter: **10 <R>**
   prompt   Specify rotation angle of text and enter: **0 <R>**
   prompt   Text and enter: **AutoCAD <R><R>**

14 Add two other text items using the following information:

	first item	*second item*
Named UCS	FRONT	RIGHT
start point	110,40,0	15,15,200
height	15	20
rotation	0	0
item	Release	2004

15 Restore UCS BASE

16 Enter paper space with **PS <R>**

17 At the command line enter **DTEXT <R>** and add the following text using the information given:
   a) start point: centred on 202.5,133.5
   b) height: 10 and rotation: 0
   c) enter text: AutoCAD <R>
      enter text: 2004 <R>
      enter text: Paper space <R><R>

18 Return to model space with **MS <R>**

19 Menu bar with **View-Hide** in each viewport and your model will be displayed with hidden line removal – Fig. 12.3(a)

20 Now select the MVLAY2 tab and:
   a) in each viewport Zoom-Extents
   b) Zoom-Scale with a 1 scale factor
   c) hide each viewport – Fig. 12.3(b)

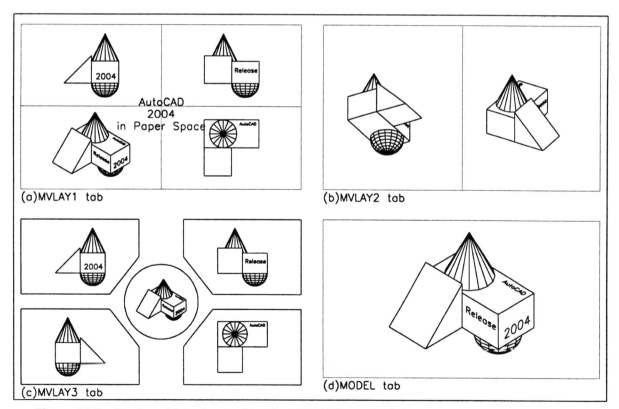

Figure 12.3 The layout tabs for checking the MV3DSTD standard sheet.

21 Select the MVLAY3 tab and repeat the three entries (a), (b) and (c) as step 20 – Fig. 12.3(c)

22 Select the Model tab to display the original 3D view of the model and hide – Fig. 12.3(d)

23 *Task*

 a) With MVLAY1 tab active and in model space, make each viewport active and menu bar with **View-Shade-Gouraud Shaded** to display the model in colour. Note the effect of the shading on the model space text.

 b) When you have completed shading, menu bar with **View-Shade-2D Wire-frame** to restore the original views of the model.

This (long) chapter is now complete. You do not have to save the drawing, but if you do be careful not to over-write your standard sheet.

We can concentrate on surface and solid modelling.

Chapter 13

Surface modelling

The best way of describing a surface model is to think of a wire-frame model with 'skins' covering all the wires from which the model is constructed. The 'skins' convert a wire-frame model into a surface model with several advantages:

1   The model can be displayed with hidden line removal

2   There is no ambiguity

3   The model can be shaded and rendered

The AutoCAD 2004 surface modeller adds **FACETED** surfaces using a polygon mesh technique, but this mesh only approximates to curved surfaces, the polygon mesh being 'planer'. The mesh density (the number of facets) is controlled by certain system variables which will be discussed in the appropriate chapter.

The different types of surface models available with AutoCAD 2004 are:

*   3D faces

*   3D meshes

*   polyface meshes

*   ruled surfaces

*   tabulated surfaces

*   revolved surfaces

*   edge surfaces

*   3D objects

Separate chapters will be used to demonstrated each surface type with worked examples.

Surface model commands can be activated:

1   From the menu bar with **Draw-Surfaces**

2   In icon form from the **Surfaces toolbar**

3   By direct keyboard entry, e.g. **3D FACE <R>**

The various exercises will use all three methods.

Chapter 14

3DFACE and PFACE

These two commands appear similar in operation, both adding faces (skins) to wire-frame models. If these added faces are in colour, the final model display can be quite impressive.

3DFACE example

A 3DFACE is a three or four sided surface added to an object and is independent of the UCS position.

1 Close any existing drawings then menu bar with **File-New** and 'open' your **MV3DSTD** template file with layer MODEL and UCS BASE both current

2 Ensure that the Model tab is current and:
a) display toolbars to suit
b) pan the icon to the lower centre of the screen

3 Create a wire-frame model using the LINE command and the reference sizes in Fig. 14.1(a). Ensure that point 1 is (0,0,0). This is a nice exercise which should give you no problems?

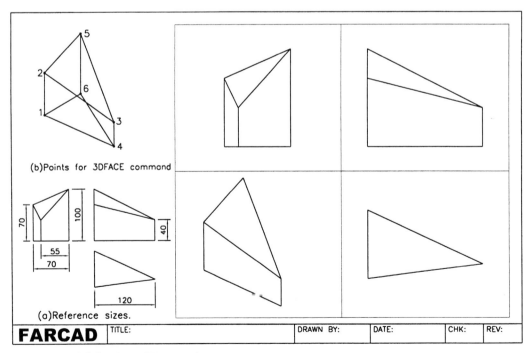

(b)Points for 3DFACE command

(a)Reference sizes.

FARCAD | TITLE: | | DRAWN BY: | DATE: | CHK: | REV:

Figure 14.1 3DFACE example.

4 The created wire-frame model has five 'planes' so make five new layers: F1 red, F2 blue, F3 green, F4 magenta and F5 colour 14

5 Make layer F1 current

6 Still with Model tab active, menu bar with **Draw-Surfaces-3DFACE** and:
prompt   Specify first point
respond   **Endpoint icon and pick pt1**
prompt   Specify second point
respond   **Endpoint icon and pick pt2**
prompt   Specify third point
respond   **Endpoint icon and pick pt3**
prompt   Specify fourth point
respond   **Endpoint icon and pick pt4**
prompt   Specify third point
respond   **right-click/enter**

7 Make layer F2 current and select the 3DFACE icon from the Surfaces toolbar and:
prompt   Specify first point and **pick Endpoint pt2**
prompt   Specify second point and **pick Endpoint pt3**
prompt   Specify third point and **pick Endpoint pt5**
prompt   Specify fourth point and **pick Endpoint pt2 then <R>**

8 Make layer F3 current, enter **3DFACE <R>** at the command line and:
prompt   First point and **pick Intersection pt3**
prompt   Second point and **pick Intersection pt4**
prompt   Third point and **pick Intersection pt6**
prompt   Fourth point and **pick Intersection pt2 then <R>**

9 Use the 3DFACE command and add faces to:
a) face: 1256 with layer F4 current
b) face: 146 with layer F5 current

10 Make the MVLAY tab current and **IN EACH VIEWPORT** zoom-extents then zoom to a factor of 2 (but 1.5 in the 3D view). This will centre the model in the viewports

11 In each viewport select from the menu bar:
a) **View-Shade-Flat Shaded** to give a colour effect
b) **View-Shade-2D Wire-frame** to remove the shade effect

12 Make MVLAY2 tab current then in each viewport
a) Zoom-Extents
b) Zoom to a factor of 3

13 Make MVLAY3 tab current and in each viewport zoom-extents then zoom to a factor of 3

14 Now make the MODEL tab active and:
a) Zoom-All the zoom to a factor of 1.5
b) Save the model as **MODR2004\CHEESE**

15 *Task*
a) Shade the model with the model tab active
b) Menu bar with **View-3D Orbit** and 'interactively' rotate the 3D shaded model
c) Select the Undo icon twice to restore the original 3D view

16 This first 3DFACE exercise is now complete

17 *Note*:

   a) The 3D orbit command will be discussed in detail in a later chapter. At this stage interactively rotate the model by holding down the left button and moving the pointing device

   b) In this example I have referred to the MVLAY2 and MVLAY3 tabs. In general most of the future exercises will be completed with the MVLAY1 tab active, i.e. the traditional four viewport configuration. The other layout and model tabs will occasionally be mentioned. The user should investigate these tabs in their own time

   c) Fig. 14.1 displays my 'customisation' idea for the MV3DSTD standard sheet.

Additional 3DFACE exercise

The 3DFACE command can be used to face any three or four sided 'plane'. The command allows 'continuous' faces to be created and will be demonstrated with a 2D example (although the procedure is valid in 3D) so:

1 Begin any New metric drawing from scratch and refer to Fig. 14.2

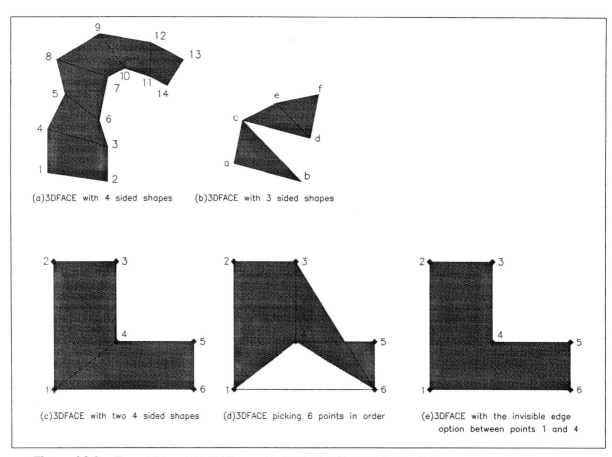

Figure 14.2 The additional 3DFACE exercise (with Flat Shaded, Edges On).

2   Activate the 3DFACE command and:
    prompt   Specify first point or [Invisible]
    respond   **pick any pt1 on the screen**
    prompt   Specify second point or [Invisible]
    respond   **pick any pt2 on the screen**
    prompt   Specify third point or [Invisible]<exit>
    respond   **pick any pt3** (which is the 1st pt of the next face)
    prompt   Specify fourth point or [Invisible]<create three-sided
              face>
    respond   **pick any pt4** (which is the 2nd pt of the next face)
    and   Face 1-2-3-4 displayed
    prompt   Specify third point
    respond   **pick any pt5** (3rd pt of face and 1st pt of next face)
    prompt   Specify fourth point
    respond   **pick any pt6** (4th pt of face and 2nd pt of next face)
    and   Face 3-4-5-6 displayed
    prompt   Specify third point
    respond   **pick any pt7** (3rd pt of face and 1st pt of next face)
    prompt   Specify fourth point
    respond   **pick any pt8** (4th pt of face and 2nd pt of next face)
    and   Face 5-6-7-8 displayed
    prompt   Specify third point..
    respond   in response to the third and fourth point prompts:
           a) pick any points 9 and 10
           b) then pick any points 11 and 12
           c) then pick any points 13 and 14 in response
           d) then right-click/enter
    and   Faces 7-8-9-10; 9-10-11-12; 11-12-13-14 will be displayed
           as Fig. 14.2(a)

3   Activate the 3DFACE command again and:
    prompt   Specify first point and pick any pta
    prompt   Specify second point and pick any ptb
    prompt   Specify third point and pick any ptc (1st of next face)
    prompt   Specify fourth point and press <RETURN> – note the prompt
    and   Face a-b-c displayed
    prompt   Specify third point and pick any ptd (2nd pt of face)
    prompt   Specify fourth point and pick any pte (3rd pt of face)
    and   Face c-d-e displayed
    prompt   Specify third point and pick any ptf
    prompt   Specify fourth point and <R><R> – two returns
    and   Face d-e-f displayed as Fig. 14.2(b)

4   Now erase the two continuous 3D faces.

The invisible 3DFACE edge

When the 3DFACE command is used to create continuous three/four sided 'shapes', all the sides of the face are displayed. It is possible to create a 3DFACE with an 'invisible edge' which will be demonstrated with the following example.

1   Still referring to Fig. 14.2

2   Draw any L-shaped object as Fig. 14.2(c) and copy it to two other parts of the screen

3 Set the running object snap to ENDPOINT and make a new layer with any name and colour to suit. Make this layer current

4 *a*) activate the 3DFACE command and using the first L shape:
   prompt   Specify first point and pick pt1
   prompt   Specify second point and pick pt2
   prompt   Specify third point and pick pt3
   prompt   Specify fourth point and pick pt4 then right-click/enter
   b) activate the 3DFACE command and:
   prompt   Specify first point and pick pt1
   prompt   Specify second point and pick pt6
   prompt   Specify third point and pick pt5
   prompt   Specify fourth point and pick pt4 then right-click/enter

5 Two 3D faces have been created, with two edges between points 1 and 4. Why two edges?

6 With the second L shape, activate the 3DFACE command and:
   prompt   Specify first point and pick pt1
   prompt   Specify second point and pick pt2
   prompt   Specify third point and pick pt3
   prompt   Specify fourth point and pick pt4
   prompt   Specify third point and pick pt5
   prompt   Specify fourth point and pick pt6
   prompt   Specify third point and right-click/enter
   and   the 3DFACE effect is displayed as Fig. 14.2(d) with an edge between points 1 and 4 and between points 3 and 6. Any idea why?

7 *a*) with the third L shape, 3DFACE again and:
   prompt   Specify first point and pick pt1
   prompt   Specify second point and pick pt2
   prompt   Specify third point and pick pt3
   prompt   Specify fourth point and enter: **I <R>**
   prompt   Specify fourth point and pick pt4
   prompt   Specify third point and right-click/enter
   and   the 3DFACE is displayed without edge 1–4
   b) repeat the 3DFACE command again and:
   prompt   Specify first point and pick pt1
   prompt   Specify second point and pick pt6
   prompt   Specify third point and pick pt5
   prompt   Specify fourth point and enter: **I <R>**
   prompt   Specify fourth point and pick pt4
   prompt   Specify third point and right-click/enter
   and   the 3DFACE is displayed without edge 1–4

8 The third L shape now displays two 3D faces, without an edge between points 1 and 4 as Fig. 14.2(e)

9 *a*) Menu bar with **View-Shade-Flat Shaded**
   b) Menu bar with **View-Shade-Flat Shaded, Edges On** – Fig. 14.2
   c) Restore the layout to its original display with **View-Shade-2D Wire-frame**

10 This exercise is now complete. It can be saved if required, but it will not be used again

11 *Note*:
   While a 2D example has been used to demonstrate the invisible edge, the same procedure is used to add a 3DFACE to a 3D model.

PFACE

A PFACE is a polygon mesh.

It is similar to a 3DFACE. It allows the user to define a number of vertices for the surface to be faced, not the 3 or 4 allowed with the 3DFACE command. Using the command requires excessive keyboard entry, as the following example will demonstrate.

1   Open your MV3DSTD template file and:

   a) MVLAY1 tab active

   b) make four new layers: F1 blue, F2 green, F3 magenta, F4 cyan

   c) layer MODEL current, restore UCS FRONT

   d) model space with the lower left viewport (3D) active

2   Set the elevation to 0 and the thickness to −200

3   Select the POLYGON icon from the Draw toolbar and:

   a) number of sides: 5

   b) centre of polygon: 0,0

   c) circumscribed circle radius: 50

4   Zoom-extents then zoom to a factor of 1 in all viewports

5   Refer to Fig. 14.3 which only displays the 3D viewport of this exercise

6   Menu bar with **View-Hide** to display the pentagonal prism without a 'front vertical surface' – Fig. 14.3(a)

7   Set the object snap to ENDPOINT, make layer F1 current and:

   a) select the 3DFACE icon and:

First point	pick pt2
Second point	pick pt3
Third point	pick pt4
Fourth point	<R><R> – two returns

   b) select the 3DFACE icon and:

First point	pick pt2
Second point	pick pt6
Third point	pick pt5
Fourth point	pick pt4
Third point	right-click/enter

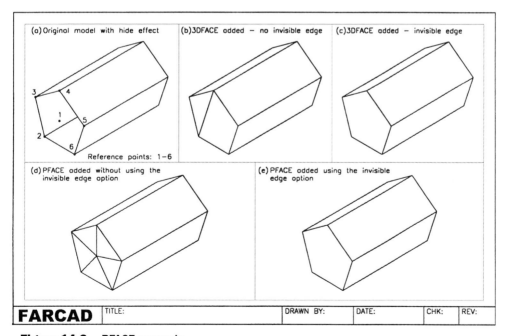

Figure 14.3   PFACE example.

8 Hide the model to display the model as Fig. 14.3(b) with hidden line removal. Restore the original wire-model from the menu bar with View-Regen

9 Make layer F2 current and freeze layer F1 then:
   a) activate the 3DFACE command and:

First point	pick pt2
Second point	pick pt3
Third point	enter **I <R>** then pick pt4
Fourth point	<R><R> – two returns

   b) 3DFACE again and:

First point	pick pt2
Second point	pick pt6
Third point	pick pt5
Fourth point	enter **I <R>** then pick pt4
Third point	right-click/enter

10 Hide to display the model with the 3DFACE invisible edge effect as Fig. 14.3(c)

11 Make layer F3 current and freeze layer F2

12 At the command line enter **PFACE <R>** and:

prompt	Specify location for vertex 1 and enter: **0,0 <R>**, i.e. pt1
prompt	Specify location for vertex 2 and pick pt2
prompt	Specify location for vertex 3 and pick pt3
prompt	Specify location for vertex 4 and pick pt4
prompt	Specify location for vertex 5 and pick pt5
prompt	Specify location for vertex 6 and pick pt6
prompt	Specify location for vertex 7 and right-click as no more vertices
prompt	Face 1,vertex 1
then	Enter a vertex number or [Color/Layer] and enter: **1 <R>**
and	Enter the following in response to the vertex number prompts:
prompt	Face 1, vertex 2 and enter: **2 <R>**
prompt	Face 1, vertex 3 and enter: **3 <R>**
prompt	Face 1, vertex 4 and <RETURN> i.e. end of face 1
prompt	Face 2, vertex 1 and enter: **1**
prompt	Face 2, vertex 2 and enter: **3**
prompt	Face 2, vertex 3 and enter: **4**
prompt	Face 2, vertex 4 and <RETURN> i.e. end of face 2
prompt	Face 3, vertex 1 and enter: **1**
prompt	Face 3, vertex 2 and enter: **4**
prompt	Face 3, vertex 3 and enter: **5**
prompt	Face 3, vertex 4 and <RETURN> i.e. end of face 3
prompt	Face 4, vertex 1 and enter: **1**
prompt	Face 4, vertex 2 and enter: **5**
prompt	Face 4, vertex 3 and enter: **6**
prompt	Face 4, vertex 4 and <RETURN> i.e. end of face 4
prompt	Face 5, vertex 1 and enter: **1**
prompt	Face 5, vertex 2 and enter: **6**
prompt	Face 5, vertex 3 and enter: **2**
prompt	Face 5, vertex 4 and <RETURN> i.e. end of face 5
prompt	Face 6, vertex 1 and <RETURN> to end command

13 Menu bar with View-Hide to display the end of the prism with a PFACE surface – Fig. 14.3(d)

14 Make layer F4 current and freeze layer F3

15 Repeat the PFACE command line entry and:
    prompt   Specify location for vertex 1 and enter: **0,0 <R>**
    prompt   Specify location for vertex 2 and pick pt2
    prompt   Specify location for vertex 3 and pick pt3
    prompt   Specify location for vertex 4 and pick pt4
    prompt   Specify location for vertex 5 and pick pt5
    prompt   Specify location for vertex 6 and pick pt6
    prompt   Specify location for vertex 7 and <RETURN>
    prompt   Face 1, vertex 1, Enter a vertex number and enter: **−1 <R>**
    prompt   Face 1, vertex 2 and enter: **2 <R>**
    prompt   Face 1, vertex 3 and enter: **−3 <R>**
    prompt   Face 1, vertex 4 and <RETURN>
    prompt   Face 2, vertex 1 and enter: **−1**
    prompt   Face 2, vertex 2 and enter: **3**
    prompt   Face 2, vertex 3 and enter: **−4**
    prompt   Face 2, vertex 4 and <RETURN>
    prompt   Face 3, vertex 1 and enter: **−1**
    prompt   Face 3, vertex 2 and enter: **4**
    prompt   Face 3, vertex 3 and enter: **−5**
    prompt   Face 3, vertex 4 and <RETURN>
    prompt   Face 4, vertex 1 and enter: **−1**
    prompt   Face 4, vertex 2 and enter: **5**
    prompt   Face 4, vertex 3 and enter: **−6**
    prompt   Face 4, vertex 4 and <RETURN>
    prompt   Face 5, vertex 1 and enter: **−1**
    prompt   Face 5, vertex 2 and enter: **6**
    prompt   Face 5, vertex 3 and enter: **−2**
    prompt   Face 5, vertex 4 and <RETURN>
    prompt   Face 6, vertex 1 and <RETURN>

16 Now View-Hide to display the model with the hidden edge option of the PFACE command

17 The negative (−1, etc.) entry for the face/vertex prompt is the invisible edge option. Think this out for the vertices being entered

18 Save if required, but this model will not be used again.

Summary

A 3DFACE is a planer surface 'added' to three or four sided planes and is independent of the UCS position.

1 The 3DFACE and PFACE commands allow **surface models** to be created by drawing 'skins' over wire-frame models

2 The HIDE command allows surface models to be displayed with hidden line removal

3 Although the 3DFACE command can only be used with three or four sided 'shapes' although continuous faces can be created

4 The PFACE command can be used with multi-sided figures

5 Both commands have an invisible edge option

6   The commands can be activated:
    a) 3DFACE: command line, menu bar or icon
    b) PFACE: command line only

7   Using separate coloured layers for different faces allows models to be displayed in colour using the SHADE command

8   If the HIDE command has been used, the original model can be restored with:
    a) View-Regen: in the active viewport
    b) View-Regenall for all viewports

9   If the SHADE command has been used then REGEN will not restore the original model. The menu bar sequence View-Shade-2D Wire-frame must be selected. This must be applied in the viewports which used the SHADE command.

Assignment

One of MACFARAMUS's most famous structures was a temple created from a series of hexagonal shaped columns. The prisms used for these columns had both a horizontal and sloped surface. It is one of these columns which you have to create as a 3DFACED surface model and then array the surface model in a circular layout.

Activity 9: Hexagonal column of MACFARAMUS.

1   Start with your MV3DSTD template/drawing file

2   MVLAY1 tab, layer MODEL current and UCS BASE current

3   Create a 3D wire-frame model of the hexagonal prism using the sizes given

4   Make three new coloured layers for:
    a) vertical sides: blue
    b) slopes: green
    c) horizontal surfaces: red

5   Use the 3DFACE command with the coloured layers to convert the wire-frame model into a surface model. Note that two of the vertical sides have five vertices and you should attempt to use the invisible edge option with these two sides. The base has (of course) six sides and again, you should try to 3DFACE the base using the invisible edge option.

6   When all 3D faces have been added, polar array the complete surface model using the following information:
    a) objects: the complete model
    b) centre point: 0,150
    c) method: items and angle
    d) number of items: 5
    e) angle to fill: 360
    f) rotate as copied: active

7   In each viewport, zoom-extents then zoom to a factor of 1

8   Investigate the hide and shade effects on the model then save as **MODR2004\ HEXCOL**

Chapter 15

3DMESH

A 3DMESH (or more correctly, a 3D polygon mesh) consists of a series of 3D faces in a rectangular array pattern.

The actual mesh matrix is defined by **M×N vertices** where:
a) M is the number of 'columns' in the x direction
b) N is the number of 'rows' in the y direction
c) the user enters the x,y,z co-ordinates of every vertex in the matrix.

3DMESH example

1   Open your MV3DSTD template file with MVLAY1 tab, layer MODEL and UCS BASE current

2   Enter model space and make the lower right viewport active

3   Display the Surfaces toolbar and refer to Fig. 15.1

4   Zoom-centre about the point 150,150,25 at 250 magnification

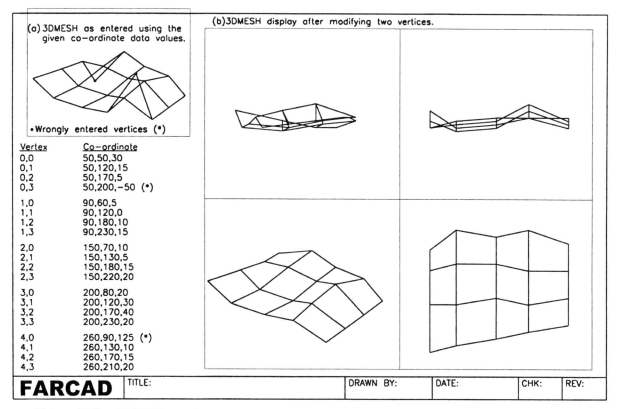

(a) 3DMESH as entered using the given co-ordinate data values.

• Wrongly entered vertices (•)

(b)3DMESH display after modifying two vertices.

Vertex	Co-ordinate
0,0	50,50,30
0,1	50,120,15
0,2	50,170,5
0,3	50,200,−50 (•)
1,0	90,60,5
1,1	90,120,0
1,2	90,180,10
1,3	90,230,15
2,0	150,70,10
2,1	150,130,5
2,2	150,180,15
2,3	150,220,20
3,0	200,80,20
3,1	200,120,30
3,2	200,170,40
3,3	200,230,20
4,0	260,90,125 (•)
4,1	260,130,10
4,2	260,170,15
4,3	260,210,20

FARCAD TITLE: DRAWN BY: DATE: CHK: REV:

Figure 15.1   3DMESH exercise.

5   Select the 3DMESH icon from the Surfaces toolbar and:
    prompt   `Enter size of mesh in M direction`
    enter   **5 <R>**
    prompt   `Enter size of mesh in N direction`
    enter   **4 <R>**
    prompt   `Specify location for vertex (0,0)`
    enter   **50,50,30 <R>**
    prompt   `Specify location for vertex (0,1)`
    enter   **50,120,15 <R>**
    prompt   `Specify location for vertex (0,2)`
    respond refer to Fig. 15.1 and enter the given co-ordinates in response to the vertex
               prompts

6   When the last vertex co-ordinate is entered – Vertex (4,3), the 3DMESH will be dis-
    played as Fig. 15.1(a)

Modifying the mesh

Two of the vertices have been entered wrongly (denoted by *), these being:
a) vertex (0,3): entered as (50,200,−50); correct co-ordinate (50,200,5)
b) vertex (4,0): entered as (260,90,125); correct co-ordinate (260,90,5)

1   To modify the created mesh, menu bar with **Modify-Object-Polyline** and:
    prompt   `Select Polyline`
    respond **pick any point on the mesh**
    prompt   `Enter an option [Edit vertex/Smooth surface/..`
    enter   **E <R>** – the edit vertex option
    prompt   `Current vertex (0,0)`
    then   `Enter an option [Next/Previous/Left/Right/..`
    and   a small X displayed at vertex (0,0)
    enter   **N <R>** repeatedly until the current vertex is (0,3)
    prompt   `Enter an option [Next/Previous/..`
    enter   **M <R>** – the move vertex option
    prompt   `Specify new location for marked vertex`
    enter   **50,200,5 <R>**: absolute co-ordinate entry for vertex (0,3)
    prompt   `Current vertex (0,3)`
    then   `Enter an option [Next/Previous..`
    enter   **N <R>** repeatedly until current vertex is (4,0)
    prompt   `Enter an option [Next/Previous/..`
    enter   **M <R>**
    prompt   `Specify new location for marked vertex`
    enter   **@0,0,−120 <R>**: relative co-ordinate entry for vertex (4,0)
    prompt   `Current vertex (4,0)`
    then   `Enter an option [Next/Previous/..`
    enter   **X <R>** – to exit edit vertex options
    prompt   `Enter an option [Edit vertex/Smooth surface/..`
    and   **right-click/enter** – to exit command

2   The mesh will be displayed as Fig. 15.1(b), with the two wrongly entered vertices
    have been 'repositioned'

3   Use the hide command in all viewports – any effect?

4   Save the drawing as the exercise is now complete.

Notes

1   This example has been a brief introduction to the 3DMESH command

2   The command requires the user to enter all vertex values as co-ordinates and is therefore very tedious to use. You can also reference existing objects if these are displayed

3   The Modify-Polyline command can be used with 3D meshes. This is the same as the 2D Modify-Polyline command

4   Selecting MESH from the 3D Objects dialogue box, will allow the user to create a 2DMESH by:
a) specifying the four corners of the mesh
b) entering the M and N sizes

5   The 3DMESH values M and N can be between 2 and 256

6   A 3DMESH is a single object

7   The command can be activated:
a) by icon selection from the Surfaces toolbar
b) from the menu bar with Draw-Surfaces
c) by entering 3DMESH at the command line.

Chapter 16

Ruled surface

A ruled surface is a polygon mesh created between two defined boundaries selected by the user.

Notes

1   The objects which can be used to define the boundaries are lines, arcs, circles, points, ellipses, 2D and 3D polylines and splines.

2   The surface created is a 'one-way' mesh of straight lines drawn between the two selected boundaries.

3   The number of straight line meshes is controlled by the system variable **SURFTAB1** which has an initial value of 6.

The ruled surface effect will be demonstrated by worked examples, the first being in 2D to allow the user to become familiar with the basic terminology.

Ruled surface Example 1

1   Begin a new 2D drawing from scratch (metric) and create two layers, MOD (red) and RULSUR (blue). Refer to Fig. 16.1

2   Display the Draw, Modify and Surfaces toolbars

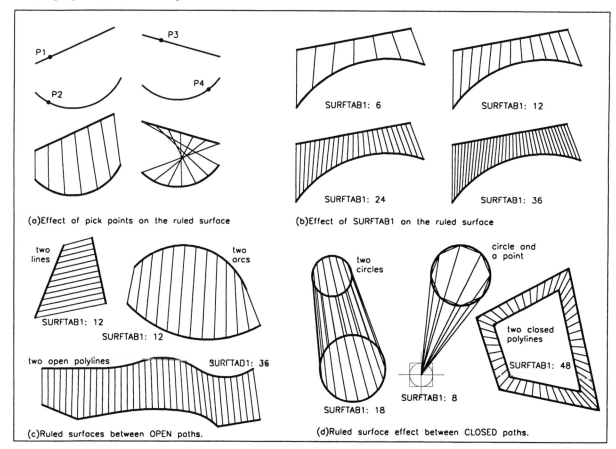

Figure 16.1   Ruled surface Example 1 – usage and basic terminology.

A. *Pick points effect*

1 With layer MOD current, draw two lines and two three point arcs as Fig. 16.1(a) then make layer RULSUR current

2 Select the RULED SURFACE icon from the Surfaces toolbar and:
prompt Select first defining curve
respond **pick a point P1 on the first line**
prompt Select second defining curve
respond **pick a point P2 on the first arc**
and a blue ruled surface is drawn between the two objects

3 Menu bar with **Draw-Surfaces-Ruled Surface** and:
a) first defining curve prompt: pick point P3 on second line
b) second defining curve prompt: pick point P4 on second arc
c) ruled surface drawn between the line and arc

4 The ruled surface drawn between selected objects is thus dependent on the pick point positions.

B. *Effect of the SURFTAB1 system variable*

1 With layer MOD current draw a line and three point arc as Fig. 16.1(b)

2 Copy the line and arc to three other places on screen

3 Make layer RULSUR current

4 At the command line enter **SURFTAB1 <R>** and:
prompt Enter new value for SURFTAB1<?>
enter **6 <R>**

5 At the command line enter **RULESURF <R>** and:
prompt Select first defining curve
respond **pick a point on the first line**
prompt Select second defining curve
respond **pick a point on the first arc**

6 By entering SURFTAB1 at the command line, enter new values of 12, 24 and 36 and add a ruled surface between the other lines and arcs.

7 *Note*:
a) The system variable **SURFTAB1** controls the display of the ruled surface effect, i.e. the number of 'strips' added between the defining curves.
b) The default value is 6.
c) The value of SURFTAB1 to be used is dependent on the 'size' of the defining curves.

C. *Open paths*

1 An open path is defined as a line, arc or open polyline

2 With layer MOD current draw some open paths as Fig. 16.1(c)

3 Using the ruled surface command and with SURFFAB1 set to your own value, add ruled surfaces between the drawn open paths.

D. *Closed paths*

1 A closed path is defined as a circle or closed polyline.

2 Draw some closed paths as Fig. 16.1(d) and add ruled surfaces between them.

E. Note

1 A ruled surface can only be drawn/added between:
   a) TWO OPEN paths
   b) TWO CLOSED paths

2 A ruled surface **cannot** be created between an open and a closed path. If a line and a circle are selected as the defining curves, the following message will be displayed:

   Cannot mix closed and open paths

3 A point can be used as a defining curve with either an open path (e.g. line) or closed path (e.g. circle)

4 The defining curves are also called boundaries

5 This first exercise is now complete and need not be saved.

Example 2

1 Open your MV3DSTD template file and refer to Fig. 16.2. Note that in Fig. 16.2 I have only displayed the 3D viewport.

2 With MVLAY1 tab, layer MODEL and UCS BASE current, zoom-centre about the point 70,40,25 at 150 magnification in all viewports

3 Create the model base from lines and trimmed circles using the sizes given in Fig. 16.2(a). Use the (0,0) start point indicated.

4 Make a new layer RULSRF, colour blue and current

5 Set the SURFTAB1 system variable to 18

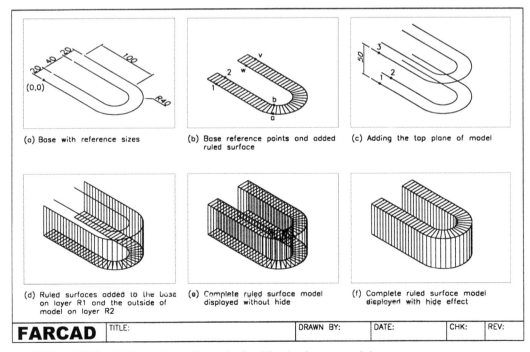

(a) Base with reference sizes

(b) Base reference points and added ruled surface

(c) Adding the top plane of model

(d) Ruled surfaces added to the base on layer R1 and the outside of model on layer R2

(e) Complete ruled surface model displayed without hide

(f) Complete ruled surface model displayed with hide effect

FARCAD	TITLE:		DRAWN BY:	DATE:	CHK:	REV:

Figure 16.2   Ruled surface Example 2 – 3D wire-frame model.

6 Using the Ruled Surface icon (three times) from the Surfaces toolbar, select the following defining curves:
   a) lines 1 and 2
   b) arcs a and b
   c) lines v and w
   d) effect as Fig. 16.2(b)

7 Erase the ruled surface and create the top surface of the model by copying the base objects:
   a) from the point 0,0,0
   b) by @0,0,50 – Fig. 16.2(c)

8 With layer RULSRF still current, select the Ruled Surface icon and select the following defining curves as Fig. 16.2(c):
   a) lines 1 and 2 – ruled surface added
   b) lines 3 and 1 – no ruled surface added and following message displayed:
      Object not usable to define ruled surface – why?
   c) Explanation:
      When the second set of defined curves was being selected:
      1. point 3 was picked satisfactorily
      2. point 1 could not be picked – you were picking the previous ruled surface added between lines 1 and 2
   d) cancel the ruled surface command (ESC) and erase the added ruled surface

9 Make the following four new layers:
   R1 – red; R2 – blue; R3 – green; R4 – magenta

10 *a*) Make layer R1 current
   b) Add a ruled surface to the base of the model (three needed)

11 *a*) Make layer R2 current
   b) Freeze layer R1
   c) Add a ruled surface to the three 'outside' vertical planes of the model
   d) Thaw layer R1 – Fig. 16.2(d)

12 *a*) Make layer R3 current
   b) Freeze layers R1 and R2
   c) Rule surface the top three defining curves of the model

13 *a*) Make layer R4 current and freeze layer R3
   b) Add a ruled surface to the three 'inside' vertical planes

14 *a*) Thaw layers R1, R2 and R3
   b) Model displayed a Fig. 16.2(e)

15 Menu bar with **View-Hide** to give Fig. 16.2(f)

16 Menu bar with **View-Shade-Gouraud Shaded** – impressive?

17 Return the model to wire-frame then save as **MODR2004\RSRF1**, it may be used in a later exercise

18 *Note*:
   When the ruled surface command is being used with adjacent surfaces, it is recommended that:
   a) a layer be made for each ruled surface to be added
   b) once a ruled surface has been added, that layer should be frozen before the next surface is added
   c) the new surface layers should be coloured for effect

19 *Task*

   a) Try the 3D orbit with the 3D viewport active

   b) The two 'ends' of the model are 'open'. A 3DFACE could be added to these ends?

   c) The original model was created from lines and circles/arcs. The base could have been created from a single polyline and then offset. Try this and add a ruled surface and note that only one set of defining curves is required. What about the SURFTAB1 value with a polyline?

Example 3

This example will investigate how a ruled surface can be added to a rectangular (square) surface which has a circular/slotted hole in it. The example will be in 2D, but the procedure is identical for a 3D model.

1 Begin a new 2D metric drawing from scratch and refer to Fig. 16.3

2 Make two new layers, MOD red (current) and RULSRF blue and set SURFTAB1 to 24

3 Using the LINE icon draw a square of side 60 with a 15 radius circle at the square 'centre' – snap on helps

4 Using the Ruled Surface icon, pick any line of the square and the circle as the defining curves. No ruled surface can be added because of the open/closed path effect – Fig. 16.3(a)

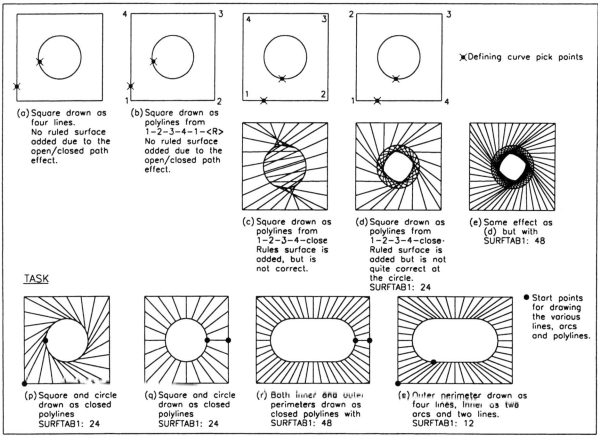

Figure 16.3 Ruled surface Example 3 – polylines and circles.

5 With the Polyline icon, draw a 60 sided square from 1–2–3–4–1–<R> as Fig. 16.3(b) and draw the 15 radius circle. Add a ruled surface and the open/closed path message is displayed and no ruled surface is added.

6 Draw a closed polyline square using the points 1–2–3–4–close in the order given in Fig. 16.3(c). Draw the circle. With the Ruled Surface icon pick the defining curves indicated and a ruled surface is added, but not as expected?

7 Draw a 60 sided square as a closed polyline and select the points 1–2–3–4–close in the order given in Fig. 16.3(d). Draw the circle then add a ruled surface picking the defining curves indicated. The added ruled surface is not quite 'correct' at the circle.

8 Erase the ruled surface effect, set SURFTAB1 to 48 and repeat the ruled surface command – Fig. 16.3(e). Set SURFTAB1 back to 24.

9 *Note*:
   a) When a ruled surface is added between two defined curves, the surface 'begins at the defined curve start points'. It is thus essential that the defined curves are:
   1 DRAWN IN THE SAME DIRECTION
   2 DRAWN FROM THE SAME 'RELATIVE' START POINT
   b) Circular holes require to be drawn as two closed polyarcs

10 *Task*
   Using the information given in step 9, add ruled surfaces to the following models displayed in Fig. 16.3:
   a) Fig. 16.3(p): square drawn as a closed polyline and circle drawn as two closed polyarcs. Note start points.
   b) Fig. 16.3(q): square drawn as a closed polyline and circle drawn as two closed polyarcs. Note that the start points differ from those in Fig. 16.3.(p).
   c) Fig. 16.3(r): both the outer and inner perimeters are drawn as closed polylines/ polyarcs. Note the start points.
   d) Fig. 16.3(s): the outer perimeter is drawn as four lines, and the inner as two arcs and two lines.

11 When this task is complete, the exercise is finished and can be saved if required. The drawing will not be used again.

Example 4

A ruled surface is one of the most effective surface modelling techniques, and I have included another 3D wire-frame model to demonstrate how it is used. The procedure when adding a ruled surface is basically the same with all models, this being:
a) create the 3D wire-frame model
b) make new coloured layers for the surfaces to be added
c) use the ruled surface command with layers current as required.

1 Open your MV3DSTD template file and refer to Fig. 16.4

2 Make four new layers, R1 red, R2 blue, R3 green and R4 magenta

3 With MVLAY1 tab and layer MODEL current, restore UCS FRONT and make the lower left (3D) viewport active

4 Select the POLYLINE icon and draw:
   Start point: 0,0
   Next point: @0,100
   Next point: **Arc option**, i.e. enter A <R>
   Arc endpoint: @50,50 then right-click/enter

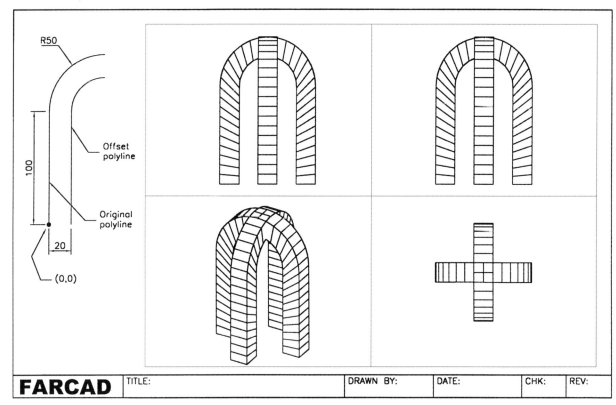

Figure 16.4 Ruled surface Example 4 – ARCHES.

5 Centre each viewport about the point 50,75,0 at 175 magnification

6 Offset the polyline by 20 'inwards'

7 Copy the two polylines from: 0,0, by: @0,0,−20

8 Change the viewpoint in the lower left viewport with the rotate option and angles:
   a) first prompt: 300
   b) second prompt: 30

9 Set SURFTAB1 to 18

10 Making each layer R1–R4 current, add a ruled surface to each 'side' of the model, remembering to freeze layers as in the second example.

11 Restore UCS BASE and polar array the complete model (crossing selection) using:
   a) Method: Total number of items & Angle to fill
   b) Centre point: X: 50 and Y: 10
   c) Total number of items: 4
   d) Angle to fill: 360
   e) Rotate items as copied: active

12 Hide, shade, etc. – impressive result?

13 Save the complete model as MODR2004\ARCHES for future recall

14 *Note*:
   The top 'square' of the arrayed arches – comments?

Summary

1   A ruled surface can be added between lines, circles, arcs, points and polylines

2   The command can be activated in icon form, from the menu bar or by keyboard entry

3   The command can be used in 2D or 3D

4   A ruled surface **CAN ONLY** be added between:
    a) two open paths, e.g. lines, arcs, polylines (not closed)
    b) two closed paths, e.g. circles, closed polylines

5   Points can be used with open and closed paths

6   With closed paths, the correct effect can only be obtained if:
    a) the paths are drawn in the same direction
    b) the paths start at the 'same relative point'

7   The system variable SURFTAB1 controls the number of ruled surface 'strips' added between the two defining curves

8   The default SURFTAB1 value is 6.

Assignment

Activity 10: Ornamental flower bed of MACFARAMUS.

MACFARAMUS designed some interesting artefacts for the famous lost city of CADOPOLIS. One of his least known creations has the 'hanging gardens' for which he made several unusual ornamental flower beds. It is one of these which you have to create as a 3D ruled surface model, the procedure being the same as in the examples:

1   Open your MV3DSTD template file, MVLAY1 tab, UCS BASE and layer MODEL active

2   Create the wire-frame model from lines and trimmed circles using the sizes given with the (0,0) start point. The vertical R50 arch requires the UCS RIGHT to be current and the R30 side curve requires UCS FRONT. Use your discretion for any sizes omitted.

3   With UCS BASE, zoom-extents then zoom to a factor of 2

4   Make four coloured layers

5   Add ruled surfaces to the 'four sides' of the model using the four new layers correctly. Use a SURFTAB1 value of 18 for most of the defining curves, but 6 for the 'side' line/ arc selection.

6   Hide, shade, 3D orbit, save

7   *Note*:
    a) I suggest that you enter paper space and zoom-window the lower left viewport then return to model space. This will make creating the wire-frame model and selecting the defining curves easier.
    b) As an alternative to (a), create the model with the MODEL tab active

Chapter 17

Tabulated surface

A tabulated surface is a parallel polygon mesh created along a path, the user defining:

a) the **path curve** – the profile of the final model

b) the **direction vector** – the 'depth' of the profile

The following are important points to note when creating a tabulated surface:

1 The path curve can be created from lines, arcs, circles, ellipses, splines or 2D/3D polylines

2 The direction vector **MUST** be a line or an open 2D/3D polyline

3 The system variable SURFTAB1 determines the 'appearance' of curved tabulated surfaces.

Example

1 Open your MV3DSTD template file with MVLAY1 tab and layer MODEL current, lower left viewport active and UCS BASE. Display toolbars to suit.

2 Refer to Fig. 17.1 (which only displays the 3D viewport) and draw two line segments:

a) *Start point*: 0,0,0 with *next point*: @0,0,120

b) *Start point*: 0,0,0 with *next point*: @−150,0,0

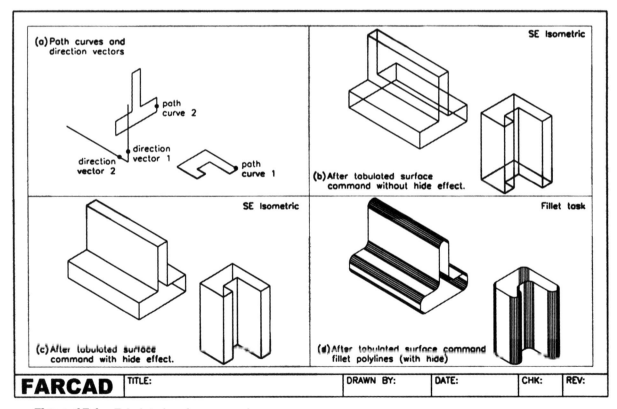

Figure 17.1 Tabulated surface example.

3 Restore the appropriate UCS and draw two closed polylines with the following co-ordinate data:

UCS BASE	*UCS RIGHT*
Start: 70,50,0	Start: 20,20,-50
Next: @50,0	Next: @100,0
Next: @0,20	Next: @0,30
Next: @-20,0	Next: @-40,0
Next: @0,30	Next: @0,80
Next: @50,0	Next: @-20,0
Next: @0,20	Next: @0,-80
Next: @-80,0	Next: @-40,0
Next: close	Next: close

4 Restore UCS BASE and zoom-centre about $-35,70,60$ at 250 magnification in all viewports

5 Select the TABULATED SURFACE icon from the Surfaces toolbar and:
 prompt Select object for path curve
 respond **pick polyline 1** as Fig. 17.1(a)
 prompt Select object for direction vector
 respond **pick line 1 at the end indicated**
 and a tabulated surface is added to the path curve

6 The added tabulated surface has a 'depth' equal to the length of the direction vector, i.e. 120

7 Menu bar with **Draw-Surfaces-Tabulated Surface** and:
 prompt Select object for path curve
 respond **pick polyline 2** as Fig. 17.1(a)
 prompt Select object for direction vector
 respond **pick line 2 at the end indicated**

8 Figure 17.1 displays (in 3D) the results of the tabulated surface operations:
 a) reference information
 b) tabulated surfaces without hide at SE Isometric viewpoint
 c) tabulated surfaces with hide at SE Isometric viewpoint

9 **Task 1**
 a) Erase the tabulated surfaces to display the original path curves
 b) Repeat the tabulated surface commands, but pick the direction vector lines at the 'opposite ends' from the exercise. The path curve will be 'extruded' in the opposite sense.
 Task 2
 a) Erase the tabulated surfaces to display the original path curves
 b) Fillet each polyline with a radius of 10 – remember how to fillet a polyline?
 c) Activate the tabulated surface command selecting the path curve and direction vertex as before
 d) The result should be as Fig. 17.1(d) with hide effect.

Summary

1 A tabulated surface is a parallel polygon mesh

2 The command requires:
 a) a path curve – a single object
 b) a direction vector – generally a line

3   The command can be used in 2D or 3D

4   The final surface orientation is dependent on the direction vector 'pick point'

5   SURFTAB1 determines the surface appearance with curved objects

6   The command can be activated:
    a) in icon form from the Surfaces toolbar
    b) from the menu bar with Draw-Surfaces
    c) by entering **TABSURF <R>** at the command line.

Chapter 18

Revolved surface

A revolved surface is a polygon mesh generated by rotating a path curve (profile) about an axis, the user selecting:
a) the **path curve** – a single object, e.g. a line, arc, circle or 2D/3D polyline
b) the **axis of revolution** – generally a line, but can be an open or closed polyline.

The generated mesh is controlled by two system variables:
a) SURFTAB1: controls the mesh in the direction of the revolution
b) SURFTAB2: defines any curved elements in the profile
c) the default value for both variables is 6.

Example 1

1 Open the MV3DSTD template file, MVLAY1 tab and layer MODEL current, UCS BASE and refer to Fig. 18.1

2 Make the lower right viewport active and display toolbars

3 Draw two lines:
   a) start point: 0,0   next point: @100,0
   b) start point: 0,0   next point: @0,100

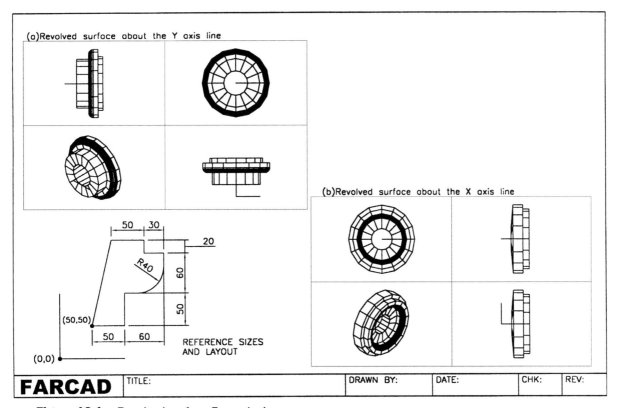

Figure 18.1   Revolved surface Example 1.

4   Set SURFTAB1 to 16 and SURFTAB2 to 6 – command line entry

5   Using the polyline icon from the Draw toolbar, create a **CLOSED** polyline shape using the reference sizes given in Fig. 18.1. The start point is to be 50,50

   Note:
   The actual polyline shape is not that important. Use your discretion/own design, but try and keep to the overall reference sizes given.

6   Select the REVOLVED SURFACE icon from the Surfaces toolbar and:
   prompt   Select object to revolve
   respond **pick any point on the polyline**
   prompt   Select object that defines the axis of revolution
   respond **pick the Y axis line**
   prompt   Specify start angle<0> and enter: **0 <R>**
   prompt   Specify included angle (+= ccw, − = cw) <360>
   enter   **360 <R>**

7   A revolved surface model will be displayed in each viewport

8   In all viewport, zoom-centre about 0,120,0 at 350 magnification

9   Hide each viewport – Fig. 18.1(a)

10 Erase the revolved surface (Regen needed?) to display the original polyline shape and from the menu bar select **Draw-Surfaces-Revolved Surface** and:
   a) object to revolve: pick the polyline shape
   b) object to define axis of revolution: pick the X axis line
   c) start angle: 0
   d) included angle: 360

11 Zoom-centre about 100,0,0 at 400 magnification

12 Hide the viewports – Fig. 18.1(b)

13 *Task*
   a) Gouraud shade the 3D viewport
   b) Menu bar with **View-3D Orbit** and interactively rotate the shaded model
14 Save if required, as this first exercise is complete.

Example 2

1   Open the MV3DSTD template file, MVLAY1 tab, layer MODEL, UCS BASE with the lower right viewport active

2   Refer to Fig. 18.2

3   Draw a line from 0,0 to @0,250

4   With the polyline icon, draw an **OPEN** polyline shape using the sizes in Fig. 18.2(a) as a reference. The start point is to be (0,50) but the final polyline shape is at your discretion – it is your wine glass design.

5   Set SURFTAB1 to 18 and SURFTAB2 to 6

6   At the command line enter **REVSURF <R>** and:
   a) object to revolve: pick the polyline shape
   b) object to define axis of revolution: pick the line
   c) start angle: enter 0
   d) included angle: enter 270

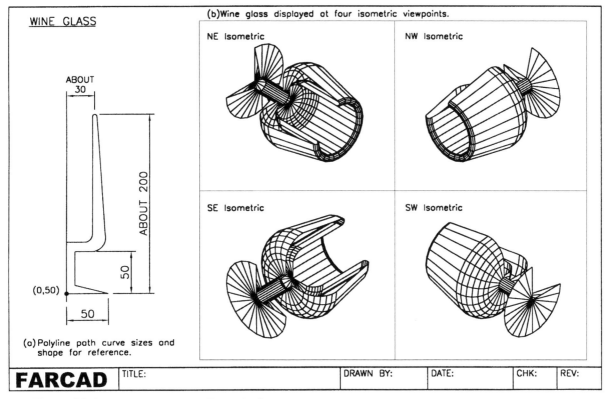

WINE GLASS

(a) Polyline path curve sizes and shape for reference.

(b) Wine glass displayed at four isometric viewpoints.

NE Isometric   NW Isometric

SE Isometric   SW Isometric

FARCAD	TITLE:		DRAWN BY:	DATE:	CHK:	REV:

Figure 18.2   Revolved surface Example 2.

7   Set the following 3D viewpoints in the named viewports:
Top left: NE Isometric   Top right: NW Isometric
Lower left: SE Isometric   Lower right: SW Isometric

8   Zoom-centre about 0,120,0 at 200 magnification

9   Hide the viewports – Fig. 18.2(b)

10 With the model tab active, Gouraud shade and 3D orbit the model

11 The exercise is complete and can be saved if required.

Summary

1   The revolved surface command can be used to produce very complex surface models from relatively simple profiles.

2   The resultant polygon mesh is controlled by the two system variables SURFTAB1 and SURFTAB2:
a) SURFTAB1: controls the mesh in the direction of rotation
b) SURFTAB2: controls the display of curved elements in the profile

3   The start angle can vary between 0 and 360. A start angle of 0 means that that the surface is to begin on the current drawing plane. This is generally what is required.

4   The included angle allows the user to define the angle the path curve is to be revolved through. The 360 default value gives a complete revolution, but 'cut-away' models can be obtained with angles less than 360.

5 The direction of the revolved surface is controlled by the sign of the included angle and:
  a) +ve for anti-clockwise revolved surfaces
  b) −ve for a clockwise revolved surface

6 The command can be activated by icon, from the menu bar or by command line entry.

Assignment

MACFARAMUS designed a garden furniture arrangement for the gardens in CADOPOLIS. This garden furniture set complemented the ornamental flower bed created as a ruled surface. You need to create two profiles and revolve them about two different axes. Adding colour to the revolved surfaces greatly enhances the model appearance with shading and rendering.

Activity 11: Garden furniture set of MACFARAMUS.

1 Use your MV3DSTD template file – MVLAY1 tab

2 Make the top right viewport active and restore UCS FRONT

3 Draw two polyline profiles using the reference data given. Use your discretion for sizes not given, or design your own table and chair. Also draw two vertical lines for the axes of revolution.

4 Set SURFTAB1 to 18 and SURFTAB2 to 6

5 Revolve the profiles about vertical lines

6 Change the colour of the revolved chair to green and the table to blue

7 Restore UCS BASE and make the lower left viewport active

8 Polar array the chair for five items about the point (0,0) with rotation

9 Hide the viewports, then REGENALL and save as MODR2004\GARDEN

10 Try the following with the Model tab active:
  a) Gouraud shade
  b) Use the 3D orbit command

11 Make sure this model has been saved.

Chapter 19

Edge surface

An edge surface is a 3D polygon mesh stretched between four **touching** edges. The edges can be combinations of lines, arcs, polylines or splines but **must form a closed loop**.

The edge surface mesh is controlled by the system variables:
a) SURFTAB1: the M facets in the direction of the first edge selected
b) SURFTAB2: the N facets in the direction of the edges adjacent to the first selected edge

Three examples will be used to demonstrate the command, the first being in 2D, the second to allow us to use the editing features of a polygon mesh and the third will use splines as the four touching edges.

Example 1 (2D edge surfaces)

1   Open any 2D drawing and make two layers, EDGE colour red and MESH colour blue

2   Refer to Fig. 19.1 and display toolbars as required

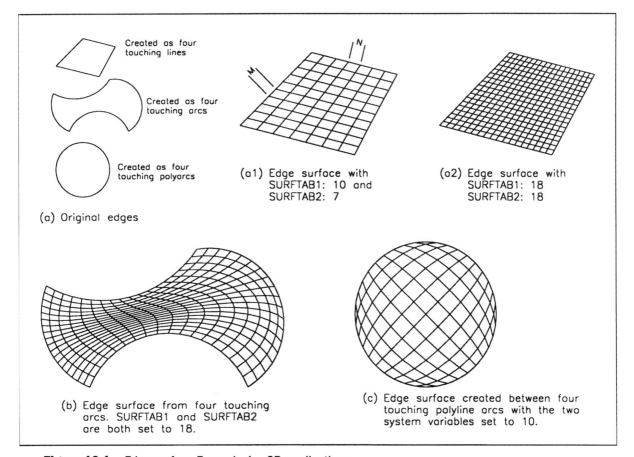

Figure 19.1   Edge surface Example 1 – 2D application.

3 With layer EDGE current, create the following touching edges similar in layout to Fig. 19.1(a):
   a) four lines
   b) four three point arcs
   c) four single 90 degree polyline arcs (use CE option)

4 Set SURFTAB1 to 10 and SURFTAB2 to 7 (command line entry)

5 Select the EDGE SURFACE icon from the Surfaces toolbar and:
   prompt   Select object 1 for surface edge
   respond pick a point on any line
   prompt   Select object 2 for surface edge
   respond pick a point on another line
   prompt   Select object 3 for surface edge
   respond pick a point on a third line
   prompt   Select object 4 for surface edge
   respond pick a point on the fourth line

6 A 10 × 7 surface mesh is stretched between the four touching lines as Fig. 19.1(a1)

7 *a*) Erase the added edge surface
   b) Set SURFTAB1 and SURFTAB2 to 18
   c) Menu bar with **Draw-Surfaces-Edge Surface** and pick the four touching lines in any order
   d) The edge surface mesh is displayed as Fig. 19.1(a2)

8 At the command line enter **EDGESURF <R>** and pick the four arcs to display the edge surface mesh as Fig. 19.1(b)

9 Set both SURFTAB1 and SURFTAB2 to 10 and add an edge surface mesh between the four touching polyarcs – Fig. 19.1(c). The result of this mesh is quite interesting?

   This completes the first exercise and it need not be saved.

Example 2 (a 3D edge surface mesh)

1 Open your MV3DSTD template file (MVLAY1 tab)

2 Refer to Fig. 19.2

3 With layer MODEL, UCS BASE and the lower left viewport active, use the LINE icon to draw the four touching lines:
   Start point: 0,0,0
   Next point: 150,0,−20
   Next point: 180,200,30
   Next point: 40,120,50
   Next point: close

4 The four lines will be displayed as Fig. 19.2(a)

5 Centre each viewport about the point 90,100,25 at 250 magnification

6 Make a new layer, MESH colour blue and current

7 Set both SURFTAB1 and SURFTAB2 to 10

8 Using the edge surface icon, pick the four lines in the order indicated 1–2–3–4 as Fig. 19.2(a)

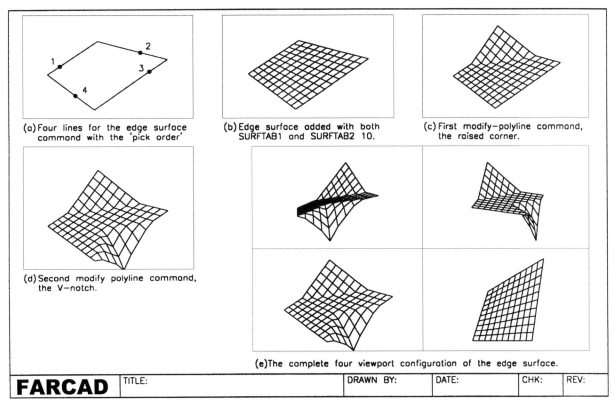

(a) Four lines for the edge surface command with the 'pick order'

(b) Edge surface added with both SURFTAB1 and SURFTAB2 10.

(c) First modify–polyline command, the raised corner.

(d) Second modify polyline command, the V–notch.

(e) The complete four viewport configuration of the edge surface.

FARCAD TITLE: DRAWN BY: DATE: CHK: REV:

Figure 19.2 Edge surface Example 2 – 3D with edit vertices.

9   An edge surface mesh will be stretched between the four lines as Fig. 19.2(b)

10   In paper space, zoom the 3D viewport and return to model space

11   Menu bar with **Modify-Object-Polyline** and:
prompt   Select polyline
respond   **pick any point on mesh**
prompt   Enter an option [Edit vertex/Smooth surface/..
enter   **E <R>** – the edit vertex option
prompt   Current vertex (0,0)
then   Enter an option [Next/Previous/..
and   an X is displayed at the 0,0 vertex – leftmost?
enter   **U <R>** until X is at vertex (10,0) and:
prompt   Current vertex (10,0)
then   Enter an option [Next/Previous/..
enter   **M <R>** – the move option
prompt   Specify new location for marked vertex
enter   **@0,0,60 <R>**
and   **DO NOT EXIT COMMAND**

12   We now want to alter other vertices of the mesh to create a raised effect. This will be achieved by:
    a) moving the X to the required vertices
    b) using the M option
    c) entering the required relative vertex co-ordinates

13 Use the N/D/L/R/U options and enter the following new locations for the named vertices:

relative movement	vertices					
@0,0,50	9,0	10,1				
@0,0,40	8,0	9,1	10,2			
@0,0,30	7,0	8,1	9,2	10,3		
@0,0,20	6,0	7,1	8,2	9,3	10,4	
@0,0,10	5,0	6,1	7,2	8,3	9,4	10,5

14 When all the new vertex locations have been entered:
a) enter X <R> to exit the edit vertex option
b) then enter X <R> to end the command

15 The mesh will be displayed with a 'raised corner' as Fig. 19.2(c)

16 At the command line enter **PEDIT <R>** then:
a) pick any point on the mesh
b) enter E <R> for the edit vertex option
c) use the N/U/R/L/D entries to move the X to the following named vertices and with the M (move option), enter the following new locations:

relative movement	vertices					
@0,0,−80	4,10					
@0,0,−50	3,10	4,9	5,10			
@0,0,−30	2,10	3,9	4,8	5,9	6,10	
@0,0,−10	1,10	2,9	3,8	4,7	5,8	6,9	7,10

d) exit the vertex option with X <R>
e) exit the polyline edit command with X <R>

17 These vertex modifications have produced a v-type notch in the mesh as Fig. 19.2(d)

18 Paper space and zoom previous then model space. Freeze the MODEL layer

19 The complete four viewport configuration of the edge surface mesh is displayed in Fig. 19.2(e)

20 The exercise is now complete – save if required

21 Now investigate the effect of the smooth surface option (S) on the mesh.

Example 3 (an edge surface mesh created from splines)

This example will demonstrate how an edge surface can be stretched between four spline curves to simulate a car body panel.

1 Open your MV3DSTD template file as usual, i.e. MVLAY1 tab, layer MODEL and UCS BASE

2 In all viewports, zoom-extents then zoom to a factor of 1.75, but 1.5 in the 3D viewport

3 With the lower left viewport active, refer to Fig. 19.3 and by selecting the SPLINE icon from the Draw toolbar, draw four spline curves using the following co-ordinate information:

	spline 1	spline 2	spline 3	spline 4
first point	0,0,0	−200,0,0	−200,120,120	0,0,0
next point	200,0,0	−200,0,100	0,100,90	0,0,75
next point	<RETURN>	−200,120,120	<RETURN>	0,100,90
next point	–	<RETURN>	–	<RETURN>
start tan	0,0,0	−200,0,0	−200,120,120	0,0,0
end tan	−200,0,0	−200,120,120	0,100,90	0,100,90

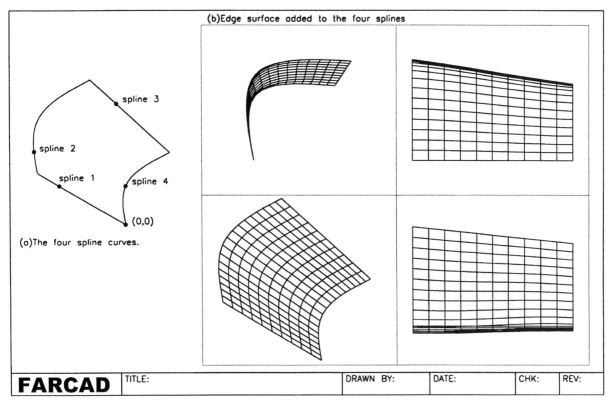

Figure 19.3 showing (b)Edge surface added to the four splines, FARCAD title block, and (a)The four spline curves with spline 1, spline 2, spline 3, spline 4 and (0,0).

Figure 19.3   Edge surface Example 3 – using splines.

4   The four spline curves will be displayed as Fig. 19.3(a)

5   Make a new layer, MESH colour blue and current

6   Set both SURFTAB1 and SURFTAB2 to 18

7   Using the Edge Surface icon, pick the four spline curves and the result should be similar to Fig. 19.3(b)

8   *Task*
   a) Use the View-Hide sequence in the four viewports, and you will probably not notice any difference
   b) Gouraud shade the four viewports and the effect is impressive
   c) Use the 3D orbit command in the 3D viewport, and real-time rotate the shaded model. This is really impressive.

9   Save if required, as this completes the edge surface exercises.

Summary

1   An edge surface is a polygon mesh stretched between four **touching** objects – lines, arcs, splines or polylines

2   An edge surface mesh can be edited with the polyline edit command

3   The added surface is a **COONS** patch and is **bicubic,** i.e. one curve is defined in the mesh M direction and the other is defined in the mesh N direction

4   The first curve (edge) selected determines the mesh M direction and the adjoining curves define the mesh N direction

5   The mesh density is controlled by the system variables:
    a) SURFTAB1: in the mesh M direction
    b) SURFTAB2: in the mesh N direction

6   The default value for SURFTAB1 and SURFTAB2 is 6

7   The type of mesh stretched between the four curves is controlled by the **SURFTYPE** system variable and:
    a) SURFTYPE 5 – Quadratic B-spline
    b) SURFTYPE 6 – Cubic B-spline (default)
    c) SURFTYPE 8 – Bezier curve

8   The SURFTYPE variable controls the appearance of all mesh curves.

Assignment

Activity 12: The flat-topped hill made by MACFARAMUS.

MACFARAMUS was contracted by the emperor TOOTENCADUM to create a flat-topped hill for a future project.

The activity is very similar to the second example, i.e. an edge surface has to have several of its vertices modified to give a 'flat-top hill' effect. The process is quite tedious, but persevere with it as it is needed for another activity in a later chapter.

1   Use your MV3DSTD template file with UCS BASE as usual

2   Zoom-centre about 0,0,50 at 400 magnification originally

3   With layer MODEL, create four touching polyline arcs of radius 200 with 0,0 as the arc centre point. If you are unsure of this, use the Centre, Start, End ARC option.

4   Make a new layer called HILL, colour green and current

5   Set SURFTAB1 and SURFTAB2 to 20

6   Add an edge surface to the four touching arcs

7   I suggest that, for editing the vertices of the edge surface, you work with the model tab active at a SE Isometric viewpoint

8   Use the Modify-Object-Polyline (PEDIT) command with the Edit Vertex option to move the following vertices by **@0,0,100**:

a)			6,9	6,10	6,11			
b)		7,7	7,8	7,9	7,10	7,11	7,12	7,13
c)		8,7	8,8	8,9	8,10	8,11	8,12	8,13
d)	9,6	9,7	9,8	9,9	9,10	9,11	9,12	9,13	9,14
e)	10,6	10,7	10,8	10,9	10,10	10,11	10,12	10,13	10,14
f)	11,6	11,7	11,8	11,9	11,10	11,11	11,12	11,13	11,14
g)		12,7	12,8	12,9	12,10	12,11	12,12	12,13	
h)		13,7	13,8	13,9	13,10	13,11	13,12	13,13	
i)				14,9	14,10	14,11			

9   *Note*:
    a) The named vertices all lie within a circle of radius 100
    b) Use the N/U/D/L/R entries of the edit vertex option until the named vertex is displayed then use the M option with an entry of @0,0,100

10 When all the vertices have been modified, optimise the multiple viewport viewpoints. I used four different VPOINT-ROTATE values and the effect with hide was quite 'pleasing'

11 Save this drawing as **MODR2004\HILL** as it will be used with a later activity

12 The View-Shade-Gouraud Shaded effect then 3D orbit with the model tab active is very interesting. You can rotate the model to 'see inside' the raised part of the edge surface.

13 Additional exercise
Can you edit vertices to produce a step effect as displayed in actual activity drawing. No help with this, but you have the ability to complete this.

Chapter 20

3D polyline

A 3D polyline is a continuous object created in 3D space. It is similar to a 2D polyline, but does not posses the 2D versatility, i.e. there are no variable width or arc options available with a 3D polyline. It does have limited editing options, but the real benefit of a 3D polyline is that it allows x,y,z co-ordinates to be used. A 2D polyline can only be created in the plane of the current UCS.

Example

This exercise will create a series of hill contours from splined 3D polylines, so:

1   Open your MV3DSTD template file, MVLAY1 tab with the lower left viewport active and UCS BASE

2   Refer to Fig. 20.1 and make the following new layers:
LEVEL1, LEVEL2, LEVEL3, LEVEL4 – all colour green
PATHUP: red; PATHDOWN: blue
DOWNN: magenta

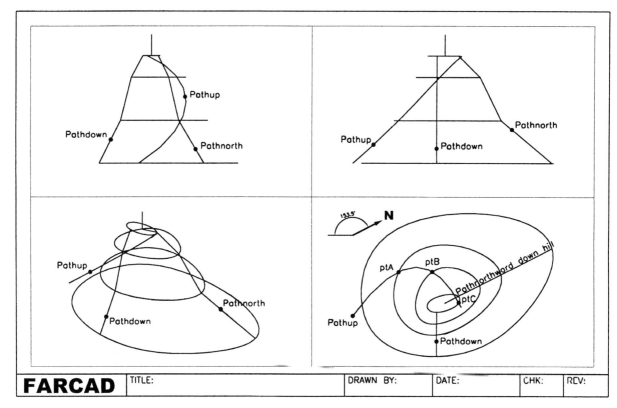

Figure 20.1   3D polyline example.

3 With layer LEVEL1 current, menu bar with **Draw-3D polyline** and:
   prompt   Specify start point of polyline and enter: **0,50,0 <R>**
   prompt   Specify endpoint of line and enter: **50,0,0 <R>**
   prompt   Specify endpoint of line and enter: **150,0,0 <R>**
   prompt   Specify endpoint of line and enter: **220,60,0 <R>**
   prompt   Specify endpoint of line and enter: **260,140,0 <R>**
   prompt   Specify endpoint of line and enter: **170,180,0 <R>**
   prompt   Specify endpoint of line and enter: **20,160,0 <R>**
   prompt   Specify endpoint of line and enter: **c <R>**

4 Make layer LEVEL2 current and at the command line enter **3DPOLY <R>** and:
   prompt   Specify start point of polyline and enter: **40,60,50 <R>**
   prompt   Specify endpoint of line and enter: **80,20,50 <R>**
   prompt   Specify endpoint of line and enter: **140,35,50 <R>**
   prompt   Specify endpoint of line and enter: **200,100,50 <R>**
   prompt   Specify endpoint of line and enter: **130,140,50 <R>**
   prompt   Specify endpoint of line and enter: **60,130,50 <R>**
   prompt   Specify endpoint of line and enter: **c <R>**

5 With layers LEVEL3 and LEVEL4 current, use the 3D polyline command with the following co-ordinate values:

	Level3	*Level4*
Start point:	70,70,100	85,70,125
Endpoint:	80,35,100	90,50,125
Endpoint:	130,45,100	130,60,125
Endpoint:	170,90,100	130,80,125
Endpoint:	100,120,100	close
Endpoint:	close	

6 Menu bar with **Modify-Object-Polyline** and:
   prompt   Select polyline
   respond **pick the 3D polyline created at level 1**
   prompt   Enter an option
   enter   **S <R>** – the Spline option
   prompt   Enter an option
   enter   **X <R>** – the exit command option

7 The selected polyline will be displayed as a splined curve

8 Use the S option of the Edit Polyline command to spline the other three 3D polylines

9 Zoom-centre about 120,90,60 at 200 magnification in all viewports

10 With layer PATHUP current, use the 3D polyline command with the following entries:
   Start: 0,50,0   –level 1 point
   Endpoint: 60,130,50   –level 2 point
   Endpoint: 100,120,100 –level 3 point
   Endpoint: 130,60,125 –level 4 point
   Endpoint: <R>

11 This 3D polyline is a path 'up the hill' and each entered co-ordinate value is a point on the level 1,2,3,4 contours

12 Spline this polyline, and it does not pass through the entered co-ordinates

13 *Task 1*

    a) Using the ID command, identify the co-ordinates of the points A, B and C where PATHUP 'crosses' the level 2, 3 and 4 contours. My values were, with UCS BASE current:

    ptA: 55.06, 101.57, 50

    ptB: 95.42, 101.91, 100

    ptC: 127.07, 65.65, 125

    b) With layer PATHDOWN current, create a 3D polyline as a 'path down the hill', this path to be drawn as a vertical line in the top (lower right) viewport. It has to 'touch' each contour, and if possible, be the shortest distance (ortho on helps, as does OSNAP nearest). Do not spline this path. Find the distance from top to bottom for this path.

    c) My value was 136.38 (remember that there are three segments)

14 *Task 2*

    MACFARAMUS erected a pole at the top of this contoured hill, probably as a signalling device. The pole was 25 units tall and its base co-ordinates were 110,65,125. From the base of this pole, a path down the hill in a northward direction was carved out of the hill. The north direction is given in Fig. 20.1. Using this information:

    a) with the DOWNN layer current, create this northward 3D polyline path down the hill, the path to just touch a contour

    b) find the distance from the base of the pole to the base of this new northward path

    c) My linear distance for the four segments of the northward path down the hill was 198.89

15 The exercise is now complete. Save if required.

Summary

1 A 3D polyline is a single object and can be used with x,y,z co-ordinate entry

2 A 3D polyline can be edited with options of Edit vertex, Spline and decurve

3 3D polylines cannot be displayed with varying width or with arc segments

4 The command is activated from the menu bar or by command line entry.

Chapter 21

3D objects

AutoCAD has nine pre-defined 3D objects, these being box, pyramid, wedge, dome, sphere, cone (and cylinder), torus, dish and mesh. They are considered as 'meshes' and can be displayed with hide, shade and render effects.

To demonstrate using some of these objects:

1   Open your MV3DSTD template file, with MVLAY1 tab, layer MODEL, UCS BASE and the lower left viewport active

2   Refer to Fig. 21.1 and display the Surfaces toolbar

3   In the steps which follow, reason out the co-ordinate entry values

4   Select the BOX icon from the Surfaces toolbar and:
prompt   Specify corner point of box and enter: **0,0,0 <R>**
prompt   Specify length of box and enter: **150 <R>**
prompt   Specify width of box and enter: **120 <R>**
prompt   Specify height of box and enter: **100 <R>**
prompt   Specify rotation angle of box about Z axis and enter: **0 <R>**

5   A red box will be displayed at the 0,0,0 origin point

6   Menu bar with **Draw-Surfaces-3D Surfaces** and:
prompt   3D Objects dialogue box
respond   **pick Wedge then OK**
prompt   Specify corner point of wedge and enter: **150,0,0 <R>**
prompt   Specify length of wedge and enter: **80 <R>**
prompt   Specify width of wedge and enter: **70 <R>**
prompt   Specify height of wedge and enter: **150 <R>**
prompt   Specify rotation angle of wedge about Z axis and enter: **−10 <R>**

7   At the command line enter **CHANGE <R>** and:
prompt   Select objects and pick the wedge then right-click
prompt   Specify change point or [Properties] and enter: **P <R>**
prompt   Enter property to change and enter: **C <R>** – colour option
prompt   Enter new colour and enter: **14 <R>**
prompt   Enter property to change and <RETURN>

8   Using the icons from the Surfaces toolbar, or the 3D Objects dialogue box, create the following two 3D objects:

Cone	*Cylinder* (using cone object)
Base centre: 50,70,100	Base centre: 75,0,50
Radius for base: 50	Radius for base: 50
Radius for top: 0	Radius for top: 50
Height: 85	Height: 90
Number of segments: 16	Number of segments: 16
Colour: green	Colour: blue

9   Restore UCS RIGHT and with the ROTATE icon from the Modify toolbar:
a) pick the blue cylinder then right-click
b) base point: 0,50
c) rotation angle: 90

10 Restore UCS BASE

11 Create another two 3D objects:

Dish	*Torus*
Centre of dish: 75,60,0	Centre of torus: 75,−90,50
Radius: 60	Radius of torus: 100
Longitudinal segments: 16	Radius of tube: 20
Latitudinal segments: 16	Tube segments: 16
Colour: magenta	Torus segments: 16
	Colour: cyan

12 With UCS RIGHT current, select the ROTATE icon:
   a) pick the cyan torus then right-click
   b) base point: −90,50
   c) angle: 90

13 Restore UCS BASE

14 Zoom centre about 75,0,50 at 300 magnification

15 Hide and shade to display the model with 'bright' colours

16 *Task*
   Still with the MVLAY1 tab active:
   a) enter paper space and make layer VP current
   b) menu bar with View-Viewports-1 Viewport and make a new viewport selecting one of the points 'outwith' the original four as shown in Fig. 21.1
   c) display the model in this new viewport from below and zoom centre about the same point as before
   d) *Question*: why did we pick one of the new viewport points outwith the original viewports?
   e) Gouraud shade the model in each viewport, then convert back to 2D wire-frame.

17 Save the layout, although we will not use it again.

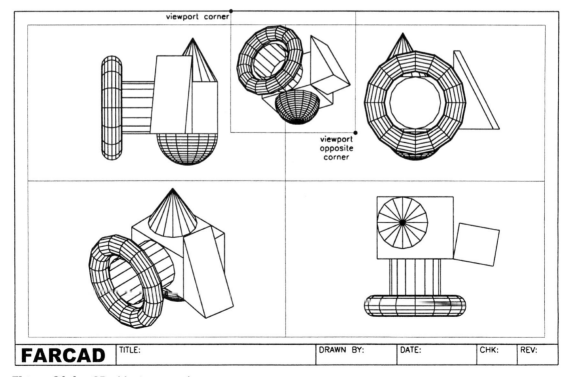

Figure 21.1 3D objects example.

Summary

1   The nine 3D objects are displayed as either faced or meshed surface models

2   The hide and shade (and render) commands can be used with 3D objects

3   3D objects are created from a reference point (corner/centre, etc.) and certain model geometry, e.g. length, width and height

4   A cylinder is created from a cone with the base and top radii having the same value

5   Activating 3D objects is available from the Surfaces toolbar or from the 3D Objects dialogue box. The various objects can also be created by command line entry, although this is **NOT** recommended.

6   *Note:*
In the 3D objects exercise, I used the command line CHANGE to alter the colour of the added objects. Changing object colour is an essential user requirement, especially when dealing with solid models. With AutoCAD 2004, there are two ways in which an objects colour can be altered, these being governed by the PICKFIRST system variable as follows:
a) PICKFIRST set to 0: CHANGE at the command line then pick the objects to be modified
b) PICKFIRST set to 1: select the objects to be modified then pick the Properties icon
c) the user must decide which method is to be used. Either method is perfectly valid
d) I will generally refer to this type of operation as:
   1. change the colour of the box to blue
   2. make the added cylinder green.

Assignment

MACFARAMUS was commissioned by the emperor TOOTENCADUM to design and build a palace for queen NEFERSAYDY in the city of CADOPOLIS. With his knowledge of CAD, MACFARAMUS decided to build the palace from 3D objects, and this is your assignment. You have to use your imagination and initiative when designing the palace. Remember that the polar array command is very useful.

Activity 13: Palace of queen NEFERSAYDY by MACFARAMUS.

1   Use your MV3DSTD template file with UCS BASE and any layout (or the model) tab

2   The 3D objects have to be positioned in a circle with an 85 **maximum** radius. This circle has to be:
a) created from four touching polyarcs – as a previous exercise
b) edge surfaced with both SURFTAB1 and SURFTAB2 set to 16. The edge surface mesh should be on its own layer with a colour number of 42.

3   The 3D objects have to be created on layer MODEL and can be to your own specification and layout. Some of my 3D objects were:

box	*wedge*	*dome*
corner: −40,−40,0	corner: 40,−40,0	centre: 0,0,60
length: 80	length: 20	radius: 25
width: 80	width: 10	colour: magenta
height: 60	height: 60	
colour: red	colour: blue	
cylinder	*cone*	
centre: 50,−50,0	centre: 50,−50,70	
radius: 8	radius: 12	
height: 70	height: 20	
colour: green	colour: green	

4   When the palace layout is complete, hide and shade

5   Save the complete model as **MODR2004\PALACE**. It will be used in a later activity.

Chapter 22

3D geometry commands

All AutoCAD commands can be used in 3D but there are three commands which are specific to 3D models, these being 3D Array, Mirror 3D and Rotate 3D. In this chapter we will investigate these three commands and how to use the Align, Extend and Trim commands with 3D models.

Getting started

To investigate the 3D commands, we will create a new model using 3D Objects, so:

1   Begin a new drawing with your MV3DSTD template file

2   For this exercise it is recommended that the user works with the Model tab active so:
    a) select the Model tab
    b) layer MODEL and UCS BASE current
    c) pan so that the UCS BASE icon is at the lower centre of the screen and alter the viewpoint with the command line entry **VPOINT <R>** then **R <R>** (rotate option) and enter angles of 300 and 30
    d) refer to Fig. 22.1 which displays the 3D viewport only

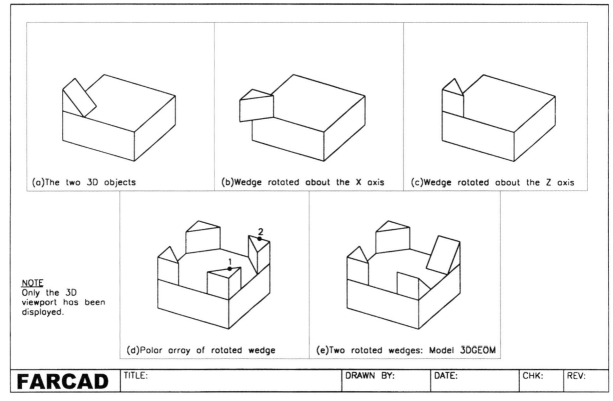

(a)The two 3D objects

(b)Wedge rotated about the X axis

(c)Wedge rotated about the Z axis

NOTE
Only the 3D viewport has been displayed.

(d)Polar array of rotated wedge

(e)Two rotated wedges: Model 3DGEOM

| **FARCAD** | TITLE: | | DRAWN BY: | DATE: | CHK: | REV: |

Figure 22.1   The 3DGEOM model for the 3D commands.

3   Select the BOX icon from the Surfaces toolbar and:
   a) corner: 0,0,0
   b) length: 100; width: 100; height: 40; Z rotation: 0
   c) colour: red

4   Select the WEDGE icon from the Surfaces toolbar and:
   a) corner: 0,0,40
   b) length: 30; width: 30; height: 30; Z rotation: 0
   c) colour: blue

5   Zoom to a scale factor of 1

6   The two 3D objects will be displayed as Fig. 22.1(a).

Rotate 3D

Using the menu bar sequence Modify-Rotate or selecting the ROTATE icon from the Modify toolbar results in a 2D command, i.e. the selected objects are rotated in the current XY plane. Objects can be rotated in 3D relative to the X, Y and Z axes with the Rotate 3D command. The command will be demonstrated by rotating the blue wedge and then the complete model.

Rotating the wedge

1   From the menu bar select **Modify-3D Operation-Rotate 3D** and:
   prompt   Select objects
   respond   **pick the blue wedge then right-click**
   prompt   Specify first point on axis or define axis by [Object/Last/..
   enter   **X <R>** – the X axis option
   prompt   Specify a point on the X axis<0,0,0>
   enter   **0,0,40 <R>** – why these co-ordinates?
   prompt   Specify rotation angle
   enter   **90 <R>**

2   The blue wedge is rotated about the x axis as Fig. 22.1(b)

3   Activate the Rotate 3D command and:
   prompt   Select objects
   enter   **pick the blue wedge the right-click**
   prompt   Rotate 3D options
   enter   **Z <R>** – the Z axis option
   prompt   Specify a point on Z axis<0,0,0>
   enter   **0,0,0 <R>**
   prompt   Specify rotation angle
   enter   **90 <R>**

4   The blue wedge is now aligned as required – Fig. 22.1(c)

5   Select the ARRAY icon from the Modify toolbar and with the Array dialogue box select:
   a) Type: Polar Array
   b) Objects: the blue wedge
   c) Centre point: X: 50 and Y: 50
   d) Method: Total number of items & Angle to fill
   e) Total items: 4
   f) Angle to fill: 360
   g) Rotate items as copied: active

6 The blue wedge is arrayed to the four corners of the box as Fig. 22.1(d)

7 With the Rotate 3D command:
   a) pick wedges 1 and 2 then right-click
   b) enter Y <R> as the defined axis
   c) enter 100,0,40 <R> as a point on axis – why this entry?
   d) enter 90 <R> as the rotation angle

8 Now move these two rotated wedges from 0,0 by @−30,0

9 The final result is Fig. 22.1(e) and can be saved as **MODR2004\3DGEOM** for future recall.

Rotating the model

1 Model 3DGEOM on the screen with model tab, UCS BASE and layer MODEL current. Refer to Fig. 22.2 (3D only displayed)

2 Menu bar with Modify-3D Operation-Rotate 3D and:
   prompt   `Select objects`
   respond **window the complete model then right-click**
   prompt   `Specify first point on axis or define axis`
   enter   **X <R>** – the X axis option
   prompt   `Specify a point on the X axis`
   respond **right-click**, i.e. accept the 0,0,0 default point
   prompt   `Specify rotation angle` and enter: **45 <R>**
   then   PAN if required then HIDE – Fig. 22.2(b)

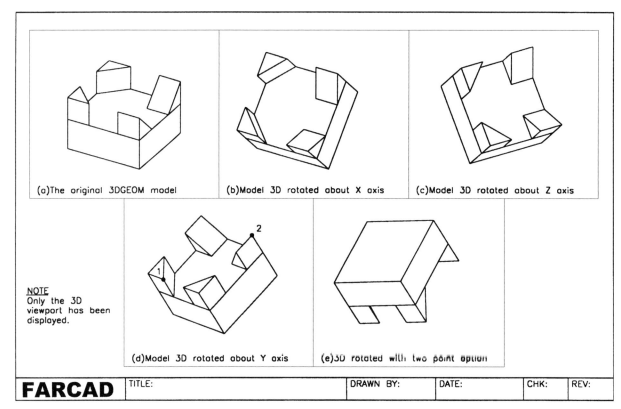

(a)The original 3DGEOM model   (b)Model 3D rotated about X axis   (c)Model 3D rotated about Z axis

NOTE
Only the 3D
viewport has been
displayed.

(d)Model 3D rotated about Y axis   (e)3D rotated with two point option

FARCAD   TITLE: DRAWN BY: DATE: CHK: REV:

Figure 22.2   The ROTATE 3D command with the 3DGEOM model.

3   At the command line enter **ROTATE3D <R>** and:
     prompt   Select objects
     respond   **window the model then right-click**
     prompt   Specify first point on axis or define axis
     enter   **Z <R>** – the Z axis option
     prompt   Specify a point on Z axis<0,0,0> and: right-click
     prompt   Specify rotation angle and enter: **60 <R>**
     then   PAN to suit and HIDE – Fig. 22.2(c)

4   Activate the Rotate 3D command and:
     a) window the model
     b) select the Y axis option
     c) accept the default 0,0,0 point on Y axis
     d) enter −60 as the rotation angle – Fig. 22.2(d) with hide

5   Rotate 3D again, window the model and:
     prompt   Specify first point on axis
     respond   **Intersection icon and pick pt1**
     prompt   Specify second point on axis
     respond   **Intersection icon and pick pt2**
     prompt   Specify rotation angle and enter: **180 <R>**
     then   PAN if needed, then HIDE – Fig. 22.2(e)

6   Shade the model then:
     a) investigate the 3D orbit effect in model space
     b) investigate the layout tabs

7   Save layout if required, but we will not use it again.

Mirror 3D

This command allows objects to be mirrored about selected points or about any of the three X–Y–Z planes.

1   Open model 3DGEOM in model tab with UCS BASE and at the 300–30 angle viewpoint

2   Refer to Fig. 22.3 which again only displays the 3D viewport

3   Menu bar with **Modify-3D Operation-Mirror 3D** and:
     prompt   Select objects
     respond   **window the model then right-click**
     prompt   Specify first point on mirror plane (3 points) or [Object/
                 Last/..
     enter   **XY <R>** – the XY plane option
     prompt   Specify point on XY plane<0,0,0> and: **right-click**
     prompt   Delete source objects<Yes/No> and enter: **Y <R>**
     then   PAN to suit – Fig. 22.3(b)

4   The model at this stage has the AMBIGUITY effect of all 3D models, i.e. are you look-
     ing down or looking up? Hence HIDE!

5 At the command line enter **MIRROR3D <R>** and:
  prompt Select objects
  respond **window the model then right-click**
  prompt Specify first point on mirror plane
  respond **Intersection icon and pick pt1**
  prompt Specify second point on mirror plane
  respond **Intersection icon and pick pt2**
  prompt Specify third point on mirror plane
  respond **Intersection icon and pick pt3**
  prompt Delete source object<Yes/No> and enter: **Y <R>**
  then   PAN to suit, then HIDE – Fig. 22.3(c)

6 Activate the Mirror 3D command and:
  a) window the model the right-click
  b) select the YZ plane option
  c) pick intersection of point A as a point on the plane
  d) enter Y to delete source objects prompt
  e) pan and hide – Fig. 22.3(d)

7 Using the Mirror 3D command:
  a) window the model the right-click
  b) select the ZX option
  c) enter −50, −50 as a point on the ZX plane
  d) accept the N default delete source objects option
  e) pan and hide – Fig. 22.3(e)

8 Shade and 3D orbit, then investigate the layout tabs

9 Save if required. We will not refer to this drawing again.

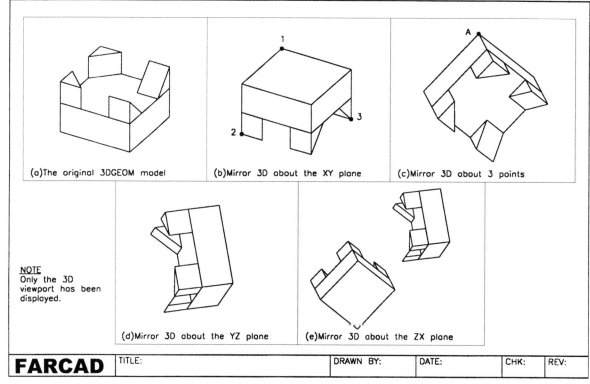

(a)The original 3DGEOM model (b)Mirror 3D about the XY plane (c)Mirror 3D about 3 points

NOTE
Only the 3D
viewport has been
displayed.

(d)Mirror 3D about the YZ plane (e)Mirror 3D about the ZX plane

FARCAD TITLE: DRAWN BY: DATE: CHK: REV:

Figure 22.3 The MIRROR 3D command with the 3DGEOM model.

3D Array

The 3D Array command is similar in operation to the 2D Array. Both rectangular and polar arrays are possible, the rectangular array having rows and columns as well as levels in the Z direction. The result of a 3D polar array requires some thought!

Rectangular

1   Open the 3DGEOM drawing, UCS BASE, Model tab active and refer to Fig. 22.4 which displays the complete MVLAY1 tab layout

2   At the command line enter **3D ARRAY <R>** and:

prompt   Select objects
respond   **window the model then right-click**
prompt   Enter the type of array [Rectangular/Polar]
enter   **R <R> – rectangular option**
prompt   Enter the number of rows (---) <1> and enter: **2 <R>**
prompt   Enter the number of columns (|||) <1> and enter: **3 <R>**
prompt   Enter the number of levels (...) <1> and enter: **4 <R>**
prompt   Specify the distance between rows (---) and enter: **120 <R>**
prompt   Specify the distance between columns (|||) and enter: **120 <R>**
prompt   Specify the distance between levels (...) and enter: **100 <R>**

3   The model will be displayed in a 2 × 3 × 4 rectangular matrix pattern but will 'be off the screen'.

4   Zoom-extents then hide to display the complete array

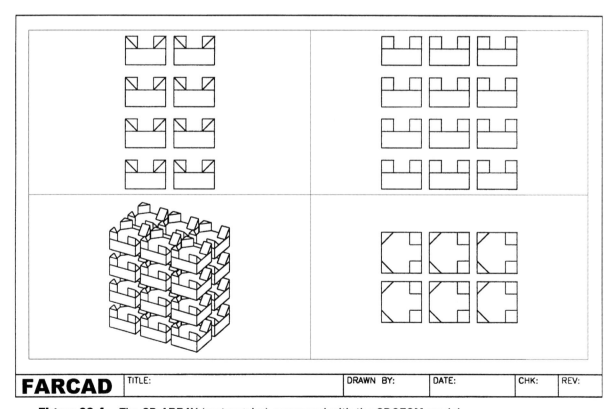

FARCAD	TITLE:		DRAWN BY:	DATE:	CHK:	REV:

Figure 22.4   The 3D ARRAY (rectangular) command with the 3DGEOM model.

5   With MVLAY1 tab active, zoom-centre about the point 170,110,185 (why these co-ordinates) at 400 magnification (550 in 3D view)

6   Hide the model – Fig. 22.4 then Gouraud shade the 3D viewport

7   Try the 3D orbit command with the shaded model

8   This exercise does not need to be saved.

Polar

1   Open 3DGEOM, UCS BASE with the Model tab active

2   Menu bar with **Modify-3D Operation-3D Array** and:
   prompt   Select objects
   respond   **window the model then right-click**
   prompt   Enter the type of array and enter: **P <R>**
   prompt   Enter the number of items and enter: **5 <R>**
   prompt   Specify the angle to fill and enter: **360 <R>**
   prompt   Rotate arrayed objects and enter: **Y <R>**
   prompt   Specify center point of array
   enter   **150,150,0 <R>**
   prompt   Specify second point on axis if rotation
   enter   **@0,0,100 <R>**, i.e. a vertical line

3   Zoom-all then investigate the MVLAY1 tab – Fig. 22.5(a)

4   Undo the 3D polar array

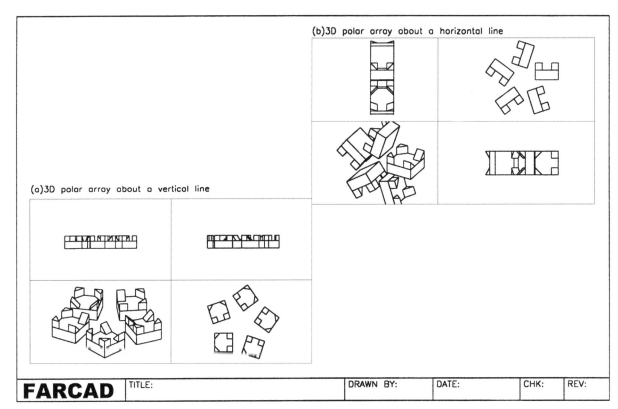

Figure 22.5   The 3D ARRAY (polar) with the 3DGEOM model.

5 Activate the 3D Array command and:
   a) window the original model then right-click
   b) enter a polar array type
   c) number of items: 5
   d) angle to fill: 360
   e) rotate as copied: Y
   f) centre point of array and enter: −50,0,0 <R>
   g) second point on axis and enter: @0,100,0 <R>

6 Zoom-all then hide and shade

7 Investigate the MVLAY1 tab which will need a zoom-extents and then a zoom-scale
   factor – Fig. 22.5(b)

8 *Note:*
   a) The first polar array was about a vertical line, the second about a horizontal line.
      Think about the entered co-ordinates
   b) The ARRAY icon/dialogue box is for 2D Arrays only

9 This exercise is complete and need not be saved.

Align

The align commands can be used in 2D or 3D and allows objects (models) to be
aligned with each other.

1 Open your MV3DSTD template file, MVLAY1 tab with UCS BASE and layer MODEL
   current. Refer to Fig. 22.6

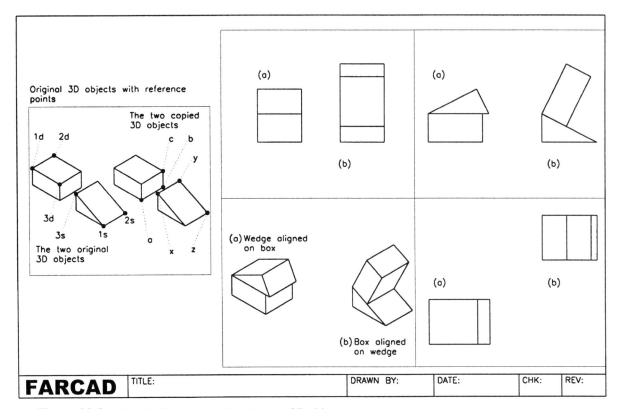

Figure 22.6 The ALIGN command using two 3D objects.

2   Using the Surfaces toolbar create the following two 3D objects:

	Box	Wedge
corner	0,0,0	160,0,0
length	100	100
wedge	80	80
height	50	50
rotation	0	0
colour	red	blue

3   Copy the box and wedge:
    a) base point: 0,0,0
    b) second point: @150,150

4   Set the running object snap to Intersection

5   Menu bar with **Modify-3D Operation-Align** and:
    prompt   Select objects
    respond   **pick the original blue wedge then right-click**
    prompt   Specify first source point and pick point 1s
    prompt   Specify first destination point and pick point 1d
    prompt   Specify second source point and pick point 2s
    prompt   Specify second destination point and pick point 2d
    prompt   Specify third source point and pick point 3s
    prompt   Specify third destination point and pick point 3d

6   The blue wedge will be aligned with it's sloped surface on the top of the box as (a)

7   At the command line enter **ALIGN <R>** and:
    a) pick the copied red box then right-click
    b) pick first source/destination points: pick a and x
    c) pick second source/destination points: pick b and y
    d) pick third source/destination points: pick c and z
    e) the red box will be aligned onto the sloped surface of the wedge as (b)

8   This completes the align exercise.

3D extend and trim

Objects can be trimmed and extended in 3D irrespective of the objects' alignment. The two commands are very dependent on the UCS position.

1   Open your MV3DSTD template file and refer to Fig. 22.7

2   Using the LINE icon draw a square with:
    Start point: 0,0
    Next point: @100,0
    Next point: @0,100
    Next point: @−100,0
    Next point: close

3   Multiple copy the square twice with:
    a) base point: 0,0
    b) second point (first copy): @0,0,120
    c) second point (second copy): @0,0,180

4   Scale the top square about the point 50,50,180 by 0.5

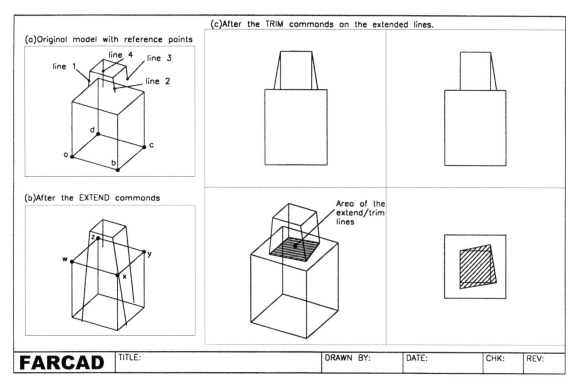

Figure 22.7 The EXTEND and TRIM commands in 3D.

5   Draw in the four vertical lines between the two large squares, then draw the following
    four lines:

line	start point	next point	colour
1	25,25,180	@0,−5,−30	blue
2	75,25,180	@5,0,−30	green
3	75,75,180	@0,5,−30	cyan
4	25,75,180	@0,0,−60	magenta

6   Change the viewpoint in the 3D viewport with VPOINT-ROTATE and angles of 300
    and 30. The model will be displayed as Fig. 22.7(a).

Extend

1   With the 3D viewport active, restore UCS RIGHT and select the EXTEND icon from
    the Modify toolbar and:
    prompt   Select objects
    respond **pick line ab then right-click**
    prompt   Select object to extend or shift-select to trim or [Project/
               Edge/Undo]
    enter   **E <R>** – the edge option
    prompt   Enter an implied edge extension mode [Extend/No extend]
    enter   **E <R>** – the extend option
    prompt   Select object to extend or shift-select to trim or [Project/
               Edge/Undo]
    enter   **P <R>** – the project option
    prompt   Enter a projection mode [None/Ucs/View]
    enter   **U <R>** – the current UCS option
    prompt   Select object to extend or shift-select to trim or [Project/
               Edge/Undo]
    respond **pick blue line 1 then right-click/enter**

2   Repeat the EXTEND command using the entries E,E,P,U as step 1 and:
    a) extend the green line 2 to edge bc with UCS RIGHT
    b) extend the cyan line 3 to edge cd with UCS FRONT

3   The extended lines are displayed as Fig. 22.7(b)

Trim

1   Still with the 3D viewport active, restore UCS BASE and select the TRIM icon from
    the modify toolbar and:
    prompt   Select objects
    respond **pick line wx then right-click**
    prompt   Select object to trim or shift-select to trim or [Project/
               Edge/Undo]
    enter   **P <R>** – the project option
    prompt   Enter a projection option [None/Ucs/View]
    enter   **V <R>** – the view option
    prompt   Select object to trim or shift-select to trim or [Project/
               Edge/Undo]
    respond *a*) make the top right viewport active
               b) pick the blue line then right-click/enter

2   Repeat the TRIM command with P and V entries as step 1 and:
    a) trim the green line to edge xy, picking the green line in the top left viewport at the
    select object to trim prompt
    b) trim the cyan line to edge yz, picking the cyan line in the top right viewport at
    the select object to trim prompt

3   The coloured lines have now been extended and trimmed 'to the top surface' of the
    large red box – Fig. 22.7(c)

Task

1   Draw four lines connecting the 'bottom ends' of the blue, green, cyan and magenta lines

2   Hatch this area

3   Find this hatched area and perimeter. My values were:
    Area: 3350; Perimeter: 232.65

Summary

1   The commands Rotate3D, Mirror3D and 3D Array are specific to 3D models.

2   Rotate 3D allows models to be rotated about the X,Y and Z axes as well as two speci-
    fied points and about objects

3   Mirror 3D allows models to be mirrored about the XY, YZ and ZX axes as well as three
    specified points and objects

4   A rectangular 3D Array is similar to the 2D command but has 'levels' in the Z direction. The result of the polar 3D Array can be difficult to 'visualise'.

5   3D models can be aligned with each other

6   The trim and extend commands can be used in 3D, the result being dependent on the UCS position. Generally objects are trimmed or extended 'to a plane'.

Chapter 23

Using blocks, wblocks and xrefs in 3D

3D blocks, wblocks and external references (xrefs) are created and inserted into a drawing in a similar manner as 2D blocks and wblocks. The UCS position and orientation is critical. In this chapter we will:

a) create a chess set using blocks
b) create a wall clock as wblocks
c) save both these models for rendering exercises
d) investigate using xrefs in 3D
e) investigate the AutoCAD Design Centre

Blocks

The original definition of an AutoCAD block was 'a group of objects saved for repetitive insertion in the drawing in which the group was created'. This definition was modified with AutoCAD 2002 and the introduction of the Design Centre, but at present we will assume the original definition to be still true.

A. Creating the models for the blocks

1 Open your MV3DSTD template file with MVLAY1 tab, layer MODEL and UCS BASE current, and the lower left viewport active.

2 In each viewport, zoom-centre about the point 120,90,50 at 275 magnification. Refer to Fig. 23.1

3 With the lower left viewport active, create the following two 3D box objects:

	box1	*box2*
corner	0,0,0	0,120,0
length	80	80
width	80	80
height	10	10
rotation	0	0
colour	number 126	number 220

4 Restore UCS FRONT and make the upper right viewport active

5 Draw a line with *start point*: 150, −10 and *next point*: @0,80

6 In paper space, zoom-window the top right viewport, then return to model space

7 *a*) Draw the pawn outline as a polyline from: 150,0 using your own design but with the 'overall' sizes given as reference
   b) Set SURFTAB1 to 16
   c) With the REVOLVED SURFACE icon from the Surfaces toolbar, revolve the pawn outline about the vertical line – full circle
   d) The pawn colour is to be red

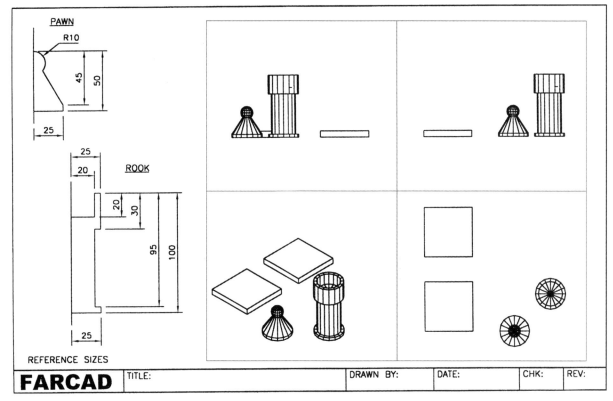

Figure 23.1 The four 3D block details.

8 Still with UCS FRONT, draw a line with *start point*: 210,−10,−60 and *next point*: @0,120

9 *a*) Draw the rook outline as a polyline from: 210,0,−60 using the reference sizes given or create your own design
   b) Revolve the rook polyline about the vertical line (360 degrees)
   c) The rook colour is to be red

10 Erase the two vertical lines

11 Restore UCS BASE and models displayed as Fig. 23.1.

B. Making the blocks

1 With the lower left viewport active and UCS BASE, menu bar with **Draw-Block-Make** and:
   prompt   Block Definition dialogue box
   respond   1. Name: enter SQ1
               2. Base point: enter X: 0; Y: 0; Z: 0
               3. pick Select objects and pick the green coloured box then right-click
               4. Objects: ensure Delete active
               5. Preview: Create icon from block geometry active
               6. Drag-and-drop units: Millimetres
               7. Description: FIRST SQUARE (Fig. 23.2)
               8. pick OK

Figure 23.2 Block definition dialogue box for SQ1.

2 The green coloured square will be 'made into a block' and should disappear from the screen due to respond 4 in step 1

3 Using the Block Definition dialogue box, make the other three blocks using the same method as step 1 with the following information:

	first	*second*	*third*
name	SQ2	PAWN	ROOK
insertion pt	0,120,0	150,0,0	210,60,0
object	purple box	3D pawn	3D rook
description	SECOND SQUARE	PAWN PIECE	ROOK PIECE

4 Now erase the original polyline outlines which should be displayed

5 Activate the block definition dialogue box and scroll at name and the four created blocks should be listed (PAWN, ROOK, SQ1, SQ2) in alphabetical order

6 *Note*:
  a) The command line BLOCK <R> entry will display the Block Definition dialogue box
  b) The command line entry – BLOCK <R> will allow block creation from the command line
  c) At the command line enter – **BLOCK <R>** and:
  prompt Enter block name or [?] and enter: **? <R>** – query option
  prompt Enter block(s) to list and enter: *** <R>**
  prompt Text window with information about the four defined blocks
  respond cancel the text window

C. *Inserting the blocks*

1 Four 'blank' viewports should be displayed with the MVLAY1 tab, UCS BASE, layer MODEL and the 3D viewport active

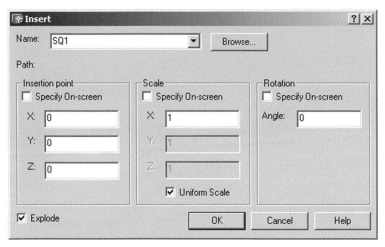

Figure 23.3 The Insert dialogue box for SQ1.

2 Menu bar with **Insert-Block** and:
   prompt Insert dialogue box
   respond 1. Name: scroll and pick SQ1
         2. Deactivate the three On-screen prompts, i.e. no tick
         3. Insertion point: enter X: 0; Y: 0; Z: 0
         4. Scale: enter X: 1
         5. Uniform Scale active, i.e. tick
         6. Rotation: enter 0 (Fig. 23.3)
         7. pick OK
   and   the green 3D box block will be displayed at the 0,0,0 insertion point

3 Repeat the Insert-Block sequence with:
   a) scroll and pick SQ2
   b) ensure on-screen prompts are off
   c) insertion point of X: 80; Y: 0; Z: 0
   d) scale of X: 1, uniform scale active and rotation: 0
   e) pick OK and the purple 3D box block will be displayed adjacent to the green 3D box

4 Zoom-centre about 320,320,100 at 700 in all viewports

5 With the 3D viewport active, complete the 64 square chessboard using one of the following methods:
   a) inserting each block SQ1 and SQ2
   b) multiple copy the two inserted boxes
   c) rectangular array
   d) note: a bit of thought needed but this operation should give you no problems

6 At the command line enter **– INSERT <R>** and:
   prompt Enter block name and enter: **PAWN <R>**
   prompt Specify insertion point and enter: **40,120,10 <R>**
   prompt Enter X scale factor and enter: **1 <R>**
   prompt Enter Y scale factor and enter: **1 <R>**
   prompt Specify rotation angle and enter: **0 <R>**
   and   the red pawn piece will be displayed on top of the left second row square

7 Now insert the ROOK full-size with 0 rotation at the insertion point of 40,40,10 using either the Insert dialogue box or the – INSERT command line entry

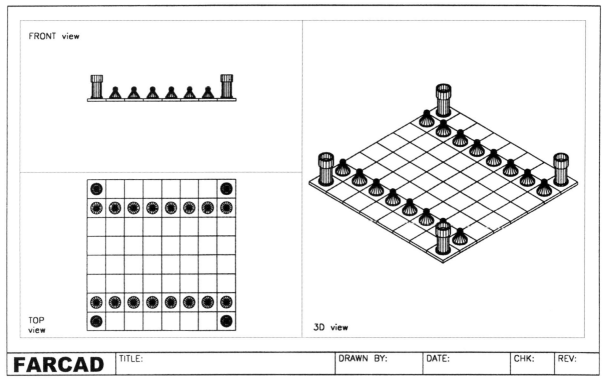

Figure 23.4   The chess layout after the block insertions.

8   *a*) Rectangular array the red pawn for 1 row and 8 columns, the column distance being 80. Row distance is 0?
    b) Copy the eight pawns from: 40,80,10 to: @0,400,0
    c) Multiple copy the rook from a base point of 40,40,10 to second points of: @560,0,0; @0,560,0; @560,560,0

9   *Task*
    a) Alter the layout to display a three viewport configuration, remembering to make layer VP current
    b) Set the viewpoints to display a 3D, top and front view
    c) Change the colour of one set of pawns and rooks to blue. This may not be as simple as you think. The inserted blocks must be exploded before the colour can be changed. Decide whether to use the PICKFIRST variable set to:
    i) 0: CHANGE at command line then pick the objects
    ii) 1: pick the objects then the Properties icon
    d) re-centre the model in each viewport about 320,320,100 at 1100 magnification in the 3D viewport and 700 in the other two

10 When complete, the layout should resemble Fig. 23.4 and should be saved as **MODR2004\CHESS** for the activity part of the chapter.

wblocks

Whereas a block can only be used in the drawing in which it was created (remember we are still assuming the original AutoCAD definition of a block), a wblock can be defined 'as a collection of objects saved for insertion into any drawing'. This means that:

Every drawing is a wblock and every wblock is a drawing

Think about this statement.

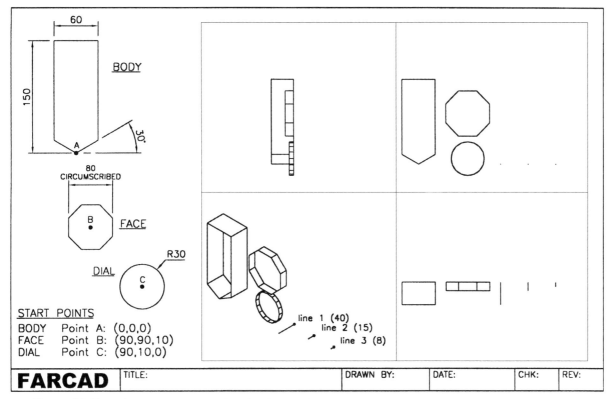

Figure 23.5   Information for creating the three wblocks.

A. Creating the models for the wblocks

1   Open the MV3DSTD template file with MVLAY1 tab, UCS BASE and zoom-centre about 125,0,100 at 300 magnification in all viewports

2   The wblocks will be created with the WCS active, so menu bar with **Tools-New UCS-World**

3   With the top right viewport active, layer MODEL current rotate the WCS about the X axis with menu bar sequence **Tools-New UCS-X** and enter 90 as the rotation angle

4   Create the three outlines for the parts of a wall clock using the following information and referring to Fig. 23.5:
    a) draw the body as lines using the 0,0,0 start point and the reference sizes given. Use the Modify-Object-Polyline command to 'convert' the five lines into a single polyline object
    b) draw the face as an octagon circumscribed in a circle with centre at 90,90,0 and radius 30
    c) draw the dial as a 30 radius circle, centre at 90,10,0

5   Draw the following three lines:
    a) line 1, *start point*: 150,0 *next point*: @0,0,40
    b) line 2, *start point*: 200,0 *next point*: @0,0,15
    c) line 3, *start point*: 250,0 *next point*: @0,0,8

6   The three wall clock components will be created as tabulated surface models, so set SURFTAB1 to 16

7   With the TABULATED SURFACE icon from the Surfaces toolbar:
   prompt   Select object for path curve
   respond   **pick the BODY polyline**
   prompt   Select object for direction vector
   respond   **pick line 1 at end indicated by the donut**
   and   extruded red tabulated surface model of wall clock body

8   Repeat the tabulated surface command and:
   a) select the FACE octagon as the path curve and line 2 as the direction vector at end indicated
   b) select the DIAL circle as the path curve and line 3 as the direction vector at end indicated

9   Change the colour of FACE surface model to blue, and the DIAL surface model to green

10   The layout at this stage should resemble Fig. 23.5

B. Making the wblocks

1   Menu bar with **Tools-New UCS-World** to restore the WCS and make the lower left viewport active

2   At the command line enter **WBLOCK <R>** and:
   prompt   Write Block dialogue box
   respond   1. Source: Objects active
                     2. Base point: X: 0; Y: 0; Z: 0
                     3. Objects: Delete from drawing active
                     4. pick Select objects and:
   prompt   Select objects at command line
   respond   **pick the red body** then right-click
   prompt   Write Block dialogue box
   respond   1. Destination file name and path and: enter **C:\MODR2004\BODY**
                     2. Insert units: Millimeters – Fig. 23.6
                     3. pick OK
   note   *a*) C:\MODR2004 is the path and folder name
                     b) BODY is the file name, i.e. the wblock name

3   A preview of the object being created into a wblock, will be displayed (briefly) at the top left of the drawing screen. The red body should have disappeared from the screen to leave the original body outline

4   Create another two wblocks using the same procedure as step 2 with the following information:

	wblock2	wblock3
Insertion base point	90,0,90	90,0,10
Objects	blue object	green object
File name and path	C:\MODR2004\FACE	C:\MODR2004\DIAL

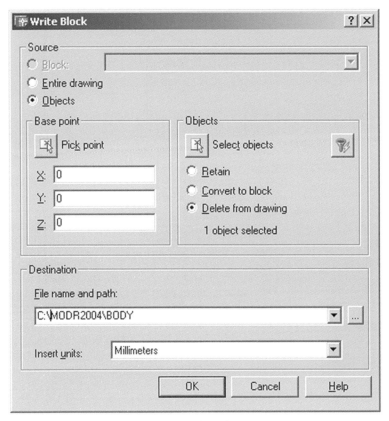

Figure 23.6 The Write Block dialogue box for BODY.

C. Inserting the three wblocks

1 Menu bar with **File-Close** and pick No to save changes

2 Menu bar with **File-Open** and 'open' your MV3DSTD template file with UCS BASE, layer MODEL, MVLAY1 tab and lower left viewport active

3 Restore the WCS

4 Menu bar with **Insert-Block** and:

   prompt   Insert dialogue box
   respond **pick Browse**
   prompt   Select Drawing File dialogue box
   respond 1. scroll and pick the C: drive
            2. scroll and pick MODR2004 (or your named folder)
            3. Scroll and pick BODY and note the preview
            4. pick Open
   prompt   Insert dialogue box
   respond 1. On-screen prompts cancelled (no ticks)
            2. Insertion point: X: 0; Y: 0; Z: 0
            3. Scale: X: 0
            4. Uniform Scale: active, i.e. tick
            5. Rotation angle: 0
            6. pick OK
   and   the red body will be displayed at the 0,0,0 insertion point

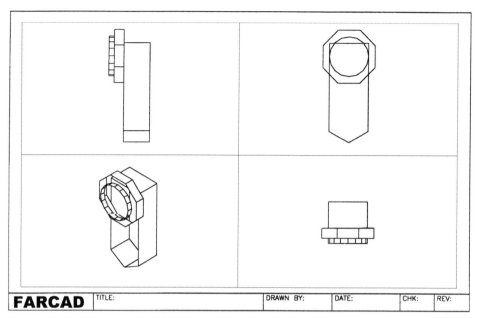

FARCAD | TITLE: | DRAWN BY: | DATE: | CHK: | REV:

Figure 23.7   The completed wall clock created from inserted wblocks.

5 Activate the Insert dialogue, select the Browse option and insert the other wblocks
   with the following information:

File name:	FACE	DIAL
Insertion point:	0,−40,130	0,−55,130
X scale:	1	1
Rotation angle:	0	0

6 In all viewports, zoom-extents then zoom to a factor of 1.5 and the drawing screen
   layout should be as Fig. 23.7

7 Investigate the other layout tabs, centring the model using the same zoom effect as
   step 6

8 With the Model tab active, set a SE Isometric viewpoint then Gouraud shade and
   investigate the 3D orbit command

9 Make the MVLAY1 tab active, then save the model as **MODR2004\WALL** as it will be
   used with the xref exercises.

xrefs

wblocks allow any objects (or complete drawings) to be inserted into any other
drawing. This is an invaluable aid to the CAD user but there is one major drawback. If
the original source (wblock) drawing is altered, the drawing into which this wblock
was inserted is not updated to reflect these modifications. This is when xrefs can
be used.

The model for the xref exercise will be the wall clock previously created. The example
is quite long as it involves opening and saving drawing files several times. This cannot
be avoided if the power of xrefs is to be fully demonstrtaed.

A. Getting ready

Open your MODR2004\WALL drawing and:
a) activate the model tab
b) ensure the WCS is current
c) zoom-centre about 0,0 at 400 magnification
d) menu bar with **Draw-Block-Base** and:
   prompt   Enter base point
   enter   **0,0,0 <R>**
e) save at this stage as **MODR2004\CLOCKDEMO**
f) close all opened drawings

B. The first layout

1   Open your MV3DSTD sheet with the Model tab active

2   Ensure the WCS is active and zoom-centre about 0,0 at 400 magnification

3   Menu bar with **Insert-External Reference** and:
   prompt   Select Reference File dialogue box
   respond   1. scroll at Look in and select your MODR2004 folder
              2. scroll the drawing names and select CLOCKDEMO
              3. pick Open
   prompt   External Reference dialogue box
   respond   1. ensure name: CLOCKDEMO
              2. Reference Type: Attachment active (block dot)
              3. Path type: Full path
              4. Cancel all On-screen prompts (no tick)
              5. Insertion point: X: 0; Y: 0; Z: 0
              6. Scale: X, Y, Z all 1
              7. Rotation: 0 (Fig. 23.8)
              8. pick OK

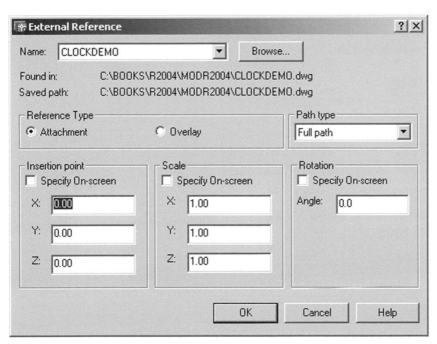

Figure 23.8   The xref dialogue box.

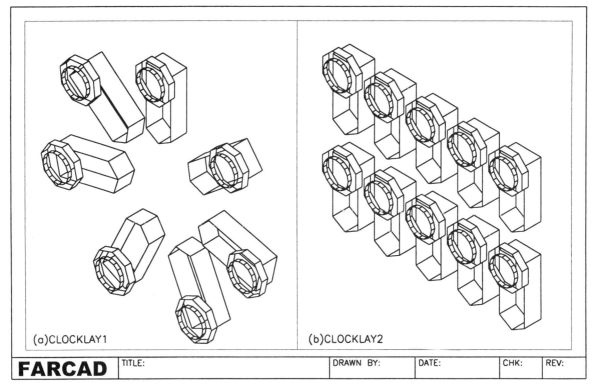

(a)CLOCKLAY1   (b)CLOCKLAY2

FARCAD | TITLE: | DRAWN BY: | DATE: | CHK: | REV:

Figure 23.9   CLOCKLAY1 and CLOCKLAY2 original layouts with xref CLOCKDEMO attached.

4 The clock model will be 'inserted' into the drawing at the WCS origin

5 *Note*:
   My path in the Fig. 23.8 dialogue box will be slightly different from your path. This is because I have saved all my work in a sub-folder (R2004) of my BOOKS folder.

6 Menu bar with **Modify-3D Operation-3D Array** and respond to the prompts with the following entries:
   a) select objects: pick the model and right-click
   b) type of array: P
   c) number of items in array: 7
   d) angle to fill: 360
   e) rotate arrayed objects: Y
   f) centre point of array: 0,0,−100
   g) second point on axis: 0,50,−100

7 Now zoom zoom-extents to give a layout similar to Fig. 23.9(a)

8 Save this layout as **MODR2004\CLOCKLAY1**

C. *The second layout*

1 Open your MV3DSTD sheet with the Model tab active

2 Ensure the WCS is active and zoom-centre about 0,0 at 400 magnification

3 Menu bar with **Insert-External Reference** and:
   prompt   Select Reference File dialogue box
   respond 1. scroll at Look in and select your MODR2004 folder
           2. scroll the drawing names and select CLOCKDEMO
           3. pick Open

prompt   External Reference dialogue box
respond 1. ensure name: CLOCKDEMO
          2. Reference Type: Attachment active (block dot)
          3. Path type: Full path
          4. Cancel all On-screen prompts (no tick)
          5. Insertion point: X: 0; Y: 0; Z: 0
          6. Scale: X, Y, Z all 1
          7. Rotation: 0
          8. pick OK

4 The clock model will be 'inserted' into the drawing at the WCS origin

5 Menu bar with **Modify-3D Operation-3D Array** and respond to the prompts with the following entries:
   a) select objects: pick the model and right-click
   b) type of array: R
   c) number of rows: 1
   d) number of columns: 5
   e) number of levels: 2
   f) column distance: 100
   g) level distance: 200

6 Now zoom zoom-extents to give a layout similar to Fig. 23.9(b)

7 Save this layout as **MODR2004\CLOCKLAY2**

D. Modifying the original CLOCKDEMO drawing

1 Open the CLOCKDEMO drawing:
   a) with WCS and model tab active
   b) with layer Model current

2 Rotate the UCS about the X axis by 90

3 Right-click LWT in the staus bar, pick Settings and:
   a) set the lineweight to 0.8
   b) pick OK

4 Draw two line segments:
   a) *Start point*: 0,0 and *next point*: @0,28
   b) *Start point*: 0,0 and *next point*: @23,0

5 The two line segments should be display with 0.8 width?

6 At the command line enter **LWDISPLAY <R>** and:
   prompt Enter new value for LWDISPLAY
   enter   **ON <R>**

7 Move the two line segments:
   a) from: 0,0
   b) to: @0,130,63

8 restore the WCS

9 Menu bar with **File-Save** to automatically update CLOCKDEMO

E. The saved clock layouts

1 Open the CLOCKLAY1 drawing to display the polar array with the clock hands in place. This drawing has been 'updated' to reflect the changes made to the CLOCK-DEMO drawing, but are the hands of the clock displayed as 'thick' lines?

2   The LWDISPLAY command must be used, so 'turn it on'

3   Press the F2 key and the AutoCAD text window will display text similar to the following:

Opening an AutoCAD 2004 format file.

*Resolve Xref 'CLOCKDEMO': C:\BOOKS\R2004\MODR2004\CLOCKDEMO.dwg 'CLOCKDEMO'
loaded.*
'CLOCKDEMO' reference file may have changed since host drawing was last saved.
Regenerating model.

AutoCAD Express Tools Copyright © 2002–2003 Autodesk, Inc.

AutoCAD menu utilities loaded.

4   The message that the reference file may have changed since the host drawing was last saved is new to R2004

5   Menu bar with File-Save to update the CLOCKLAY1 drawing

6   Open the CLOCKLAY2 drawing to display the rectangular array with the clock hands in place – remember LWDISPLAY

7   Thus when an xref is attached to a drawing, the drawing layout is automatically updated when the original xref drawing is altered

8   Menu bar with File-Save to update the layout then continue to the next section

F. Investigating the layers

1   Menu bar with Format-Layer and note the Layer Properties Manager dialogue box

2   Several new layers have been added, these being of the format:
    CLOCKDEMO/MODEL, etc.

3   There is a new CLOCKDEMO layer for every non CLOCKDEMO layer

4   The new layers have been automatically created due to the CLOCKDEMO xref being attached to the drawing, and these new layers can be 'read' as:
    a) CLOCKDEMO – the name of the attached xref
    b) | – a vertical bar symbol (commonly called a **pipe** symbol) indicating a layer with an attached xref
    c) MODEL – the actual layer name.

G. Binding an xref

When an xref (source) is attached to a drawing (destination) and this drawing is saved, the destination drawing will be automatically updated when the original source xref is modified. This is a very powerful and useful draughting aid. It is also very dangerous if not used correctly. Think about other users being able to access, and alter, all drawing data.

It may be that a drawing with an xref attached is complete and that no additional updating is required. It is then necessary to 'bind' the xref to the drawing.

AutoCAD has two bind operations, these being:
a) XBIND: an actual command which allows the user the bind specific parameters to the existing drawing, e.g. blocks, layers, text styles etc.
b) bind: an option from the Xref Manager dialogue box. Using bind will 'break the link' between the original xref and the destination drawing, and the the original xref will then become another object in the drawing.

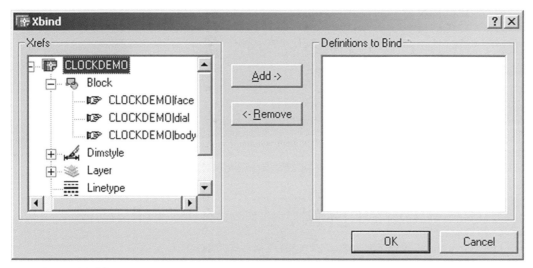

Figure 23.10 The Xbind dialogue box, expanded for CLOCKDEMO and block.

In this exercise, we will only use 'bind' but will investigate the XBIND command so:

1 CLOCKLAY2 should still be displayed

2 At the command line enter **XBIND <R>** and:
   prompt   Xbind dialogue box
   with   Xrefs on left side with CLOCKDEMO listed
   respond 1. expand CLOCKDEMO by left-click on the +
             2. expand Block by left-clicking the + (Fig. 23.10)
   and   the blocks used to make the model will be displayed

3 These individual blocks can then be bound to the drawing by selecting the block name then Add. We will not bind any individual parameters, so cancel the dialogue box.

4 Menu bar with **Insert-Xref Manager** and:
   prompt   Xref Manager dialogue box
   respond 1. pick CLOCKDEMO and it is highlighted
             2. pick Bind
   prompt   Bind Xrefs dialogue box
   respond 1. Bind Type: Bind
             2. pick OK
   prompt   Xref Manager dialogue box
   with   no attached xrefs displayed
   respond pick OK

5 Menu bar with File-Save to update CLOCKLAY2

H. Modifying the original CLOCKDEMO drawing again

1 Open the original CLOCKDEMO drawing with the hands at 3 o'clock

2 WCS should be active and LWDISPLAY should be on

3 Menu bar with **Modify-3D Operation-Rotate 3D** and:
   a) objects: select the small hand
   b) first point: 0,−63,130
   c) second point: 0,0,130
   d) rotation angle: 150

4 The clock hands are now at 8 o'clock

5 Menu bar with File-Save to update the original CLOCKDEMO

6 Now open the two clock layout drawings and:
   a) CLOCKLAY1
      1. the resolve xref message is still displayed in the text window
      2. the clock hands are at 8 o'clock, reflecting the xref which is attached to the drawing
      3. there are still CLOCKDEMO|MODEL, etc. layers
      4. XBIND still allows the CLOCKDEMO xref to be expanded
   b) CLOCKLAY2
      1. there is no resolve xref message in the text window
      2. the clock hands are still at 3 o'clock, as the original xref was 'bound' to the drawing and is not updated
      3. the pipe layers are now of the format **CLOCKDEMO0MODEL**, the vertical bar symbol having been replaced by 0 (all this is AutoCAD terminology)
      4. the XBIND command displays 'No bindable symbols present'

7 This completes the xref exercises.

The AutoCAD Design Centre

Before leaving this chapter and attempting the activities, we will investigate the Design Centre. All AutoCAD users should be aware that blocks created in a drawing can be inserted into any other drawing using the Design Centre.

1 Close all exisiting drawings then open the 3DSTDA3 standard sheet created prior to the model/paper space discussion

2 Menu bar with **Tools-Design Center** to display the Design Centre dialogue box and:
   a) position the dialogue box on the screen to suit yourself
   b) ensure that Preview and Description are active

3 In the hierarchy side (left) of the dialogue box:
   a) navigate the your named folder
   b) scroll until CHESS.dwg is displayed
   c) expand CHESS – pick the (+) at drawing icon
   d) explore Blocks, i.e. right-click on Blocks
   e) left-click PAWN from the Design Centre palette
   f) dialogue box as Fig. 23.11

4 Now right-click the SQ1 icon from the Design Centre palette and:
   prompt   Shortcut menu
   respond **pick Insert-block**
   prompt   Insert dialogue box
   with   SQ1 named and dialogue box as before
   respond pick OK and the square is inserted into the current drawing

5 This exercise is now complete. It was used to demonstrate that the Design Centre can be used to insert **ANY block from any drawing into any other drawing**

6 Close the existing drawing without saving, read the summary then attempt the two rather tricky activities.

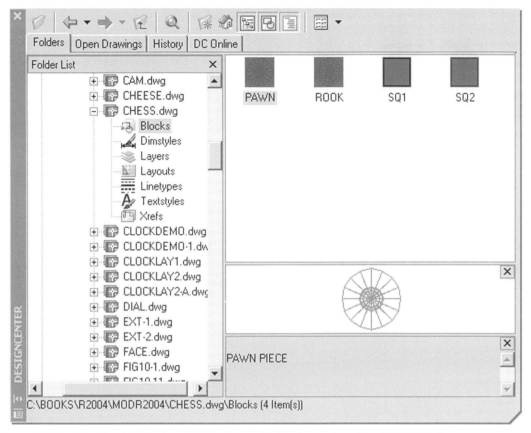

Figure 23.11   The Design Centre diologue box with blocks explored for CHESS.dwg.

Summary

1   3D blocks, wblocks and xrefs are created and inserted in a similar manner to 2D blocks, wblocks and xrefs

2   With 3D blocks, the position of the UCS is important

3   With 3D wblocks it is **strongly recommended** that the WCS be restored when creating and inserting the wblocks

4   It is also recommended that wblocks are 'stored' in the same folder as drawing file from they were created and into which they will be inserted

5   wblocks which are inserted 'unexploded' into a drawing become blocks within that drawing. It is therefore recommended that wblocks are exploded as they are inserted.

Assignments

Two activities have been included for you to attempt, one involving the partially completed chess set, and the other using two previously saved drawings, one of which will be inserted as a wblock.

Activity 14: CHESS SET

1   Recall the drawing CHESS saved earlier in this chapter to display the 64 square chessboard with the two sets of red and blue pawns and rooks

2 Design the other chess pieces – KNIGHT, BISHOP, KING and QUEEN using the same method as the worked example:

   a) draw the outline as a polyline

   b) use the revolved surface command to create the piece as a 3D surface model

   c) the actual shape of the pieces is at your discretion

   d) ensure that your start point for the outline is known – it will be useful as the block insertion point

3 Create a block of each created piece

4 Insert the created blocks onto the chessboard

5 Complete the chess set layout, remembering to change the colours of the pieces to red and blue as appropriate

6 Save as MODR2004\CHESS

7 Investigate the various shade options with the completed model

Activity 15: Palace of Queen NEFERSAYDY built by MACFARAMUS

MACFARAMUS was last encountered building the palace for queen NEFERSAYDY. Unfortunately this palace was to be built on a flat topped hill and you have to create the layout using an existing drawing and inserting another drawing into it as a wblock.

1 Open the drawing MODR2004\HILL of the edge surface model created as Activity 12

2 Insert the wblock drawing file MODR2004\PALACE of the 3D objects created as Activity 13

3 The palace has to be positioned at the centre point of the hill top, and the co-ordinates of this point as 0,0,100. This is the only help given

4 Optimise all the layout tabs for maximum effect

5 When complete save the layout as MODR2004\HILLPAL

6 Note that I have displayed the layout at different viewpoints. This was for effect only.

Chapter 24

Dynamic viewing

Dynamic viewing is a powerful (yet underused) command which is very useful with 3D modelling as it allows models to be viewed from a perspective viewpoint. The command also allows objects to be 'cut-away' enabling the user to 'see inside' models. Dynamic viewing has it's own terminology which is obvious when you are familiar with the command, but can be confusing to new users.

The basic concept of dynamic viewing is that the user has a **CAMERA** which is positioned at a certain **DISTANCE** from the model – called the **TARGET**. The user is looking through the camera lens at the model and can **ZOOM** in/out as required. The viewing direction is from the camera lens to a **TARGET POINT** on the model. The camera can be moved relative to the stationary target, and both the camera and target can be turned relative to each other. The target can also be **TWISTED** relative to the camera. Two other concepts which the user will encounter with the dynamic view command are the **slider bar** and the **perspective icon**. The slider bar allows the user to 'scale' the variable which is current, while the perspective icon is displayed when the perspective view is 'on'.

Fig. 24.1(A) displays the various dynamic view concepts of:
a) the basic terminology
b) the slider bar
c) the perspective icon

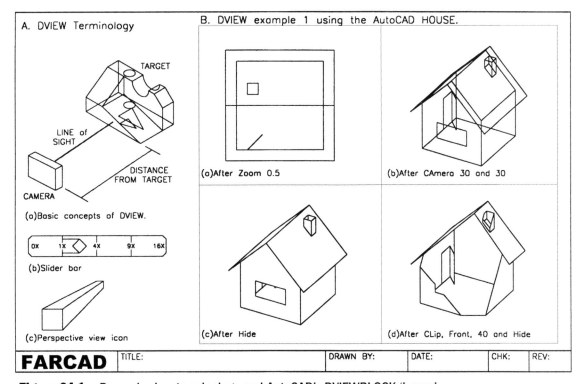

Figure 24.1 Dynamic view terminology and AutoCAD's DVIEWBLOCK 'house'.

The dynamic view command has 11 options, these being:

```
CAmera, TArget, Distance, POints, Pan, Zoom, TWist, CLip, Hide,
Off, Undo
```

The option required is activated by entering the CAPITAL letters at the command line, e.g. CA for the camera option, TW for twist, etc.

We will investigate the dynamic view command with two examples:

1 using AutoCAD's dynamic view 'house'

2 with a previously created and saved model.

Note

1 Dynamic view is a model space concept, and cannot be used in paper space

2 Dynamic view is **viewport independent**, i.e. if the command is used in a specific viewport, the model display in the other viewports will not be affected

3 The command is activated by entering **DVIEW <R>** at the command line.

Example 1 – AutoCAD's 'house'

AutoCAD has a 'drawing' – actually a type of block – which can be used as an interactive aid with the dynamic view command. We will use this house block to demonstrate some of the options so:

1 Close any existing drawings and start a new metric drawing from scratch. Refer to Fig. 24.1(B)

2 At the command line enter **DVIEW <R>** and:
   - *prompt*   Select objects or <use DVIEWBLOCK>
   - *respond* **<RETURN>**, i.e. accept the DVIEWBLOCK default
   - *a) prompt* Enter option
     [Camera/Target/Distance/Points/Pan/Zoom/Twist/Clip/
     Hide/Off/Undo]
   - *and*   some coloured lines appear on the screen
   - *enter*   **Z <R>** – the zoom option
   - *prompt* slider bar with scale displayed at top of screen
   - *and*   Specify zoom-scale factor
   - *enter*   **0.5 <R>**
   - *and*   full plan view of house – Fig. 24.1.B(a)
   - *b) prompt* Enter option [CAmera/TArget/etc
   - *enter*   **CA <R>** – the camera option
   - *prompt* *ghost image of house* which moves as mouse moved
   - *and*   Specify camera location or enter angle from XY plane
   - *enter*   **30 <R>**
   - *prompt* Specify camera location or enter angle in XY plane from
     X axis
   - *enter*   **30 <R>**
   - *and*   3D view of house – Fig. 24.1.B(b)
   - *c) prompt* Enter option [CAmera/TArget/etc
   - *enter*   **H <R>** – the hide option
   - *and*   house displayed with hidden line removal – Fig. 24. 1.B(c)

d) prompt	Enter option [CAmera/TArget/etc	
enter	**CL <R>** – the clip option	
prompt	Enter clipping option [Back/Front/Off]	
enter	**F <R>** – the front clip option	
prompt	Specify distance from target or [set to Eye(camera)/ ON/OFF]	
enter	**40 <R>**	
prompt	Enter option [CAmera/TArget/etc	
enter	**H <R>** – the hide option	
and	house displayed 'cut-away' similar to Fig. 24.1.B(d)	
e) enter	**U <R>** – undoes the hide effect of (d)	
enter	**U <R>** – undoes the clip effect of (d)	
enter	**U <R>** – undoes the hide effect of (c)	
and	**leave the house with Camera option displayed and the command prompt line options then read the explanation before proceeding**	

Explanation of the dynamic view command

Dynamic view is an **interactive** command and the various options can be used one after the other. The undo (U) option will undo the last option performed, and can be used repeatedly until all the options entered have been 'undone'. Some of the options have been used to demonstrate how the command is used, these options being zoom, camera, clip, hide and undo. The hide option is very useful as it allows the model to be displayed when other options have been entered, and removes the 'ambiguity' effect from the model. The command can be used with all 3D models, i.e. extruded, wire-frame, surface and solid. The command is also **viewport independent**, i.e. it can be used in any viewport without affecting the display in other viewports. The AutoCAD 'house' is for user-reference, and if a model is displayed on the screen, this model will assume the house orientation when the dynamic view command is completed. This will be investigated during the next example.

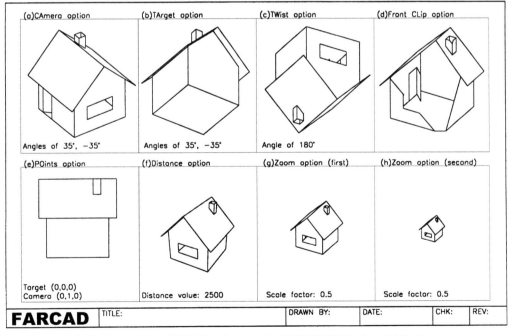

Figure 24.2 The various DVIEW options with DVIEWBLOCK – AutoCAD's house.

The house displayed on the screen has been left with the camera option with entered angles of 30 and 30. We will continue with the screen display and investigate the other dynamic view options. This means that you have to enter the various options and values as prompted.

Refer to Fig. 24.2. This drawing displays the house model orientation from one set of entered option values. The hide option has also been activated for effect.

CAmera

1   This option is used to direct the camera at the target and the camera can be 'tilted' relative to two planes with two angles:
    prompt 1 `angle in the XY plane, between` −90 degs and +90 degs
    prompt 2 `angle from the XY plane, between` −180 degs and +180 degs

2   The angles can be:
    a) toggled using the ghost image as a guide
    b) entered directly from the keyboard

3   Using the CAmera option enter the following angle values:
    angle in XY plane   *angle from XY plane*
    a)   35   35
    b)   35   −35
    c) −35   35
    d) −35   −35

4   The option can be considered similar to **VPOINT ROTATE**

5   When all the above entries have been completed, return the camera angles to the original 30 and 30, but do not exit the command.

TArget

1   This option allows the target (the model) to be tilted relative to the camera. The two angle prompts are the same as the camera option:
    prompt 1 `angle in the XY plane`
    prompt 2 `angle from the XY plane`

2   The angles can be toggled or entered from the keyboard

3   Using the TArget option enter the following angle values:
    angle in XY plane   *angle from XY plane*
    a)   35   35
    b)   35   −35
    c) −35   35
    d) −35   −35

4   The option can be used to give the same effect as the camera option, but it should be remembered that the camera and target are being 'tilted' in the 'opposite sense' to each other

5   When all angles have been entered, restore the camera to angles of 30 and 30, but do not exit the command.

TWist

1   A very useful option as it allows the 'plane' on which the target is 'resting' to be twisted through an entered angle. This angle can be positive or negative and have values between 0 and 360 degrees

2 The prompt with this option is: `Specify new view twist angle`

3 The result of the option is dependent on the CAmera/TArget angles

4 Using the TWist option enter angle values of:
   a) 35   *b*) −35   *c*) 180   *d*) −90

5 This is one of the few AutoCAD commands which allows models to be 'flipped' over by 180 degrees

6 When the four twist angles have been viewed with the hide effect, restore the original twist angle of 0, with the camera options of 30 and 30. Do not exit the command.

CLip

1 The clip option of the dynamic view command is probably the most useful of all the options, as it allows models to be 'cut-away', thus allowing the user to 'see inside' the model

2 The user selects a (F)ront or (B)ack clip and then decides on the clip distance either:
   a) using the slider bar
   b) entering a value at the command line

3 The result of the clip option is dependent on the CAmera/TArget angles as well as the 'size' of the model on the screen

4 With the CAmera angles set to 30 and 30, activate the front clip option and move the slider bar until the ghost image displays a clip effect then right-click

5 Activate the hide option then undo the hide and clip effect

6 Try some other clip option attempts then 'restore' the house at the original CAmera settings of 30 and 30.

POints

1 This option allows the model (the target) to be viewed from a specific 'stand point', the user looking at a specific point on the target

2 Two sets of co-ordinates need to be specified:
   a) the target point co-ordinates to be looked at
   b) the co-ordinates of the camera – the user

3 The co-ordinate entries can be absolute or relative

4 When this option is used, the PAn option is also usually needed

5 The result does not depend on the CAmara or TArget options

6 Use the Point option with the following entries:

	target point	*camera point*
a)	0,0,0	1,0,0
b)	0,0,0	0,1,0
c)	0,0,0	0,0,1
d)	0,0,0	1,1,0
e)	0,0,0	1,0,1
f)	0,0,0	1,1,1
g)	1,2,3	0,0,0
h)	0,0,0	1,2,3

7 The option is similar to the **VPOINT VECTOR** command

8 When all the points entries have been entered, restore the camera angles of 30 and 30, but do not exit the command.

Distance

1   Alters the distance between the camera and the target

2   The distance can be:
    a) entered as a value from the command line
    b) toggled using the slider bar

3   With the distance option, enter some values, e.g.:
    a) 1000   *b*) 1500   *c*) 2500   *d*) 5000

4   The distance option introduces **true perspective** to the model

5   When the distance option has been used, and the command is exited, the zoom command cannot be used

6   Restore the original CAmera option of 30 and 30.

Zoom

1   This option does what you would expect – it 'zooms the model'

2   The zoom factor can be:
    a) entered as a value from the keyboard
    b) toggled using the slider bar

3   Try some zoom entries, e.g.:
    a) 1   *b*) 0.75   *c*) 0.5   *d*) 0.25

4   Restore the original CAmera effect (30,30) and use the zoom option with a 0.5 scale factor three times.

Pan

1   This option is similar to the AutoCAD PAN command, but the 'real-time' pan effect is not available

2   The user selects (or enters) the pan displacement.

Hide

1   Will display the model with a hide effect

2   Removes any ambiguity.

Undo

1   Entering **U <R>** will undo the last option of the DVIEW command

2   Can be used repetitively until all the option entries have been undone.

eXit

1   Entering **X <R>** will end the dynamic view command and a blank screen will be returned

2   The blank screen is because we did not have any model displayed, the AutoCAD 'house' being a visual aid indicating what any model would 'look like'

3   With 'real models', the model orientation will be similar to the 'house' orientation when the DVIEW command has been exited, as will now be investigated

4   If the DVIEW command is still active, enter X <R>

5   This first exercise is now complete.

Example 2 – an existing 3D model

In this example we will use the dynamic view command with a previously created model.

1   Open the ruled surface model MODR2004\ARCHES created during Chapter 16 with UCS BASE, layer MODEL

2   Make the upper left viewport active and refer to Fig. 24.3

3   At the command line enter **DVIEW <R>** and:
prompt   Select objects or use <DVIEWBLOCK>
respond   **<RETURN>**
prompt   AutoCAD's house as an 'end view' in the active viewport
and   dynamic view options
enter   **CA <R>**
prompt   Angle from XY plane and enter: **−20 <R>**
prompt   Angle in XY plane and enter: **30 <R>**
prompt   dynamic view options and enter: **X <R>**

4   The model will be displayed with house CAmera configuration as Fig. 24.3(b) with hide

5   With the top right viewport active, enter **DVIEW <R>** at the command line and:
prompt   Select objects or use <DVIEWBLOCK>
respond   **window the model then right-click**
prompt   dynamic view options
enter   **TW <R>**
prompt   Specify view twist angle and enter: **−90 <R>**
prompt   dynamic view options and enter: **X <R>**
and   Model displayed with new twist as Fig. 24.3(c)

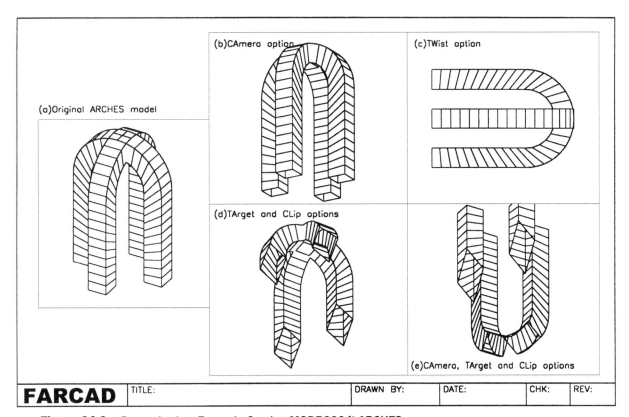

Figure 24.3   Dynamic view Example 2 using MODR2004\ARCHES.

6　Use the dynamic view command in the lower viewports with the following entries:

lower left	*lower right*
options: TArget	options: CAmera
angles: 40 and −30	angles: 30 and 30
options: Clip, Front	options: Twist, angle: 180
distance: 10	options: Clip, Front, Distance: 15
options: X	options: X
Fig. 24.3(d)	Fig. 24.3(e)

7　*Note*: if the clip-front distances do not give the exact same effect as Fig. 24.3 it could be as a result of your model having a different zoom effect from mine

8　Save the drawing if required, but we will not refer to this exercise again.

Summary

1　The dynamic view command is viewport specific, i.e. it only affects the active viewport

2　The command has several useful options:
   a) CAmera, TArget: similar to VPOINT rotate
   b) TWist: allows models to be 'inverted'
   c) Distance: introduces true perspective
   d) CLip: useful to 'see inside' models

3　The command can only be activated with DVIEW <R> at the command line

4　The command can be used:
   a) directly on models
   b) interactively using AutoCAD's house

5　The command is used relative to the WCS – observe the prompt line when the command is activated.

Assignment

No specific activity, but investigate DVIEW with some previously created models.

Chapter 25

Viewport specific layers

When layers are used with multiple viewports they are generally **GLOBAL**, i.e. what is drawn on a layer in one viewport will be displayed in the other viewports. This is quite acceptable for creating models but is unacceptable for certain other concepts, e.g. adding dimensions, sectioning the model, obtaining a true shape, etc. If dimensions (for example) have to be added to a model in a multi-view layout, then these dimensions should only be visible in the active viewport. This is also true for true shapes and sections.

In this chapter we will investigate how to create and use viewport specific layers by adding dimensions to an existing model.

Note

1   Remember that dimensioning is a 2D concept, the result depending on the position and orientation of the UCS. Think back to when we dimensioned the 3D wire-frame model.

2   In later chapters we will use viewport specific layers to extract sections and true shapes from solid models

3   Although dimensioning is being used to demonstrate viewport specific layers, later chapters will introduce the user to other methods of dimensioning models.

Global layers

1   Open the 3D faced model MODR2004\CHEESE from Chapter 14 and refer to Fig. 25.1

2   With the MVLAY1 tab current, UCS BASE, lower right viewport active make layer DIM current and display the Dimension toolbar

3   Select the LINEAR DIMENSION icon from the Dimension toolbar and dimension lines 1–2 and 1–3

4   Make the upper right viewport active and restore UCS FRONT

5   Linear dimension line a–b and baseline dimension lines a–c and a–d

6   The five dimensions will be displayed in all four viewports due to the GLOBAL nature of layer DIM. Figure 25.1(a) displays the five dimensions as displayed in the 3D viewport.

7   Now erase the five dimensions and restore UCS BASE

8   *Note*:
The standard sheet used to create the model had a dimension style 3DSTD which should be satisfactory for our exercises.

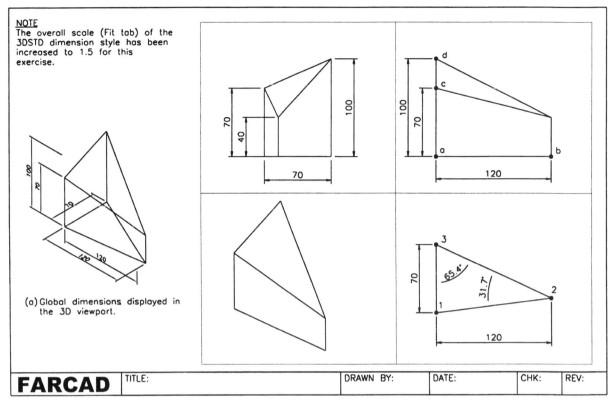

NOTE
The overall scale (Fit tab) of the 3DSTD dimension style has been increased to 1.5 for this exercise.

(a) Global dimensions displayed in the 3D viewport.

FARCAD	TITLE:		DRAWN BY:	DATE:	CHK:	REV:

Figure 25.1   Viewport specific layer exercise using MODR2004\CHEESE.

Viewport specific layers

1 Still with the 3D faced model displayed on screen?

2 Menu bar with **Format-Layer** and:

   prompt   Layer Properties Manager dialogue box
   respond 1. pick the DIM layer line – becomes highlighted
   2. pick New three times to make three new magenta layers: Layer1, Layer2 and Layer3
   3. alter the new layer names to:
      Layer1: DIMTL; Layer2: DIMTR; Layer3: DIMBR
   4. The nomenclature for these new layer names is for individual viewports, e.g. TL: top left; BR: bottom right; TR: top right
   5. pick OK

3 In model space make the top left viewport active and:
   a) activate the Layer Properties manager dialogue box
   b) select the Show Details option
   c) drag out the dialogue box to display all options
   d) hold down the control key (Ctrl) and pick the DIMTR and DIMBR layer lines
   e) pick the Freeze in Current viewport option (tick)
   f) note the Current VP icon changes from yellow to blue
   g) pick OK

4 With the top right viewport active:
   a) activate the Layer Properties Manager dialogue box
   b) by selecting the Current VP Freeze icon, freeze in the current viewports, the new layers DIMTL and DIMBR
   c) pick OK

5 With the lower right viewport active use the Layer Properties Manager dialogue box to freeze layers DIMTL and DIMTR in the current viewport

6 In the lower left viewport freeze in the current viewport, the three new layers DIMTL, DIMTR and DIMBR

7 Re-centre the model in top left, top right and lower right viewports about the point 60,35,50 at 175 magnification. This will allow additional 'space' for the dimensions

8 With UCS BASE and the lower right viewport active:
   a) make layer DIMBR current
   b) linear dimension lines 1–2 and 1–3
   c) angular dimension any two angles using the specify vertex option (snap to end-point helps)
   d) the four dimensions should only be displayed in the lower right viewport?

9 Restore UCS FRONT and make the upper right viewport active and:
   a) make layer DIMTR current
   b) linear dimension line a–b and baseline dimension lines a–c and a–d
   c) the three dimensions are only displayed in the top right viewport

10 *a*) restore UCS RIGHT
   b) make the upper left viewport active
   c) make layer DIMTL current
   d) add the four dimensions as Fig. 25.1

11 *Notes*:
   a) All dimensions should have been added to the viewport which was active when the command was used
   b) The dimensions have been added to a viewport specific layer which was made current in the viewport where the dimensions had to be added
   c) Generally the 3D viewport does not require a dimension layer as dimensions are not usually added to a 3D viewport
   d) In the 3D viewport the three viewport specific layers were all currently frozen
   e) The Overall Dimension Scale (Fit tab) of the 3DSTD dimension style was set to 1.5 for this exercise

12 The model with dimensions can now be saved with a new name.

Layer states

1 Layers can have different **states** depending on whether they are global or viewport specific, these states being easily controlled:
   a) from the Layer Properties Manager dialogue box
   b) from the icons displayed in the Object Properties toolbar

2 The different states are:

Global	*Viewport specific*
On/Off	On/Off
Freeze/Thaw	Freeze/Thaw in all viewports
Lock/Unlock	Freeze/Thaw in current/active viewports
Plot/non-plot	Freeze/Thaw in new viewports
	Lock/Unlock
	Plot/non-plot

3 Layers can have more than one state active at a time, i.e. they can:
   a) be on and locked
   b) be on and currently frozen, etc

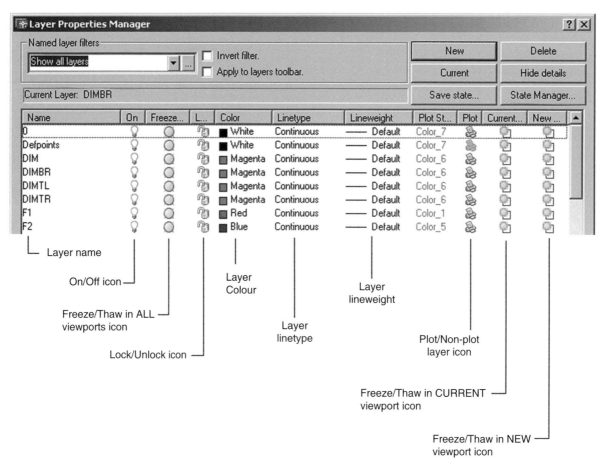

Figure 25.2   The Layer Properties Manager dialogue box.

4   Fig. 25.2 displays the Layer Properties Manager dialogue box layer icons and names and:
   a) an icon displayed in yellow is on or thawed
   b) an icon displayed in blue is off or frozen.

Paper space dimensioning

Using viewport specific layers in model space is an acceptable method of adding dimensions, but it is rather cumbersome. In later chapters, when solid modelling has been investigated, a simpler method will be discussed. There is however another method of adding dimensions to a model from paper space using the dimension variable dimension linear scale factor (**DIMLFAC**). This method is now considered 'dated' although I find it very useful. We will demonstrate the process with the same model as previous. So:

1   Re-open the original CHEESE layout – still refer to Fig. 25.1 for the reference points

2   The linear dimension from 1–2 is 120

3   In paper space, linear dimension 1–2 and the value displayed will be 77.83 – it should be 120. Dimensioning the other reference objects will also give incorrect dimension values.

4   Erase any added dimensions

5   At the command line enter **CAL <R>** and:
    prompt   Expression
    enter   **120/77.82 <R>**
    and   **1.54187** is returned at the command line
    note   this is the value of DIMLFAC

6   At the command line enter **DIMLFAC <R>** and:
    prompt   Enter new value for DIMLFAC <1.000>
    enter   **1.54187 <R>**

7   Make layer DIM current and in paper space add the linear dimensions as before. They should now display the correct values?

8   *Note*:
The value of the DIMLFAC variable depends on the 'size' of the viewports. In our example, the four viewports were all the same size, hence DIMLFAC had the same 1.54187 value. If the viewports are of different sizes then it is necessary to obtain a DIMLFAC value for each viewport before adding the paper space dimensions. The method to obtain DIMLFAC as used above:
a) model space dimension the required object
b) paper space dimension the same object
c) divide the model value by the paper value – gives DIMLFAC
d) change the DIMLFAC value
e) dimension as required.

Summary

1   Viewport specific layers are layers which are specific to a named viewport

2   Viewport specific layers are used with multi-view drawings and are essential for such concepts as dimensioning and extracting details from solid models

3   Viewport specific layers can **ONLY BE CREATED** if the TILEMODE variable is set to 0, i.e. if paper space is active

4   Viewport specific layers can be created using the:
a) Layer Properties Manager dialogue box – recommended
b) VPLAYER command – not considered in this exercise

5   The most commonly used state with viewports specific layers is 'Freeze in current viewport'

6   Using the DIMLFAC variable allows paper space dimensions to be added to a multi-view layout.

Assignment

Activity 16: Dimensioning a 3D model.

No specific assignment has been set for this chapter, but the user should try and dimension some previously created models. I have included two examples in Activity 16, these being:
a) the 3D ruled surface flower bed from Activity 10
b) the 3D faced hexagonal column from Activity 9

You have to decide on whether to add dimension using:
a) model space viewport specific layers
b) paper space with the DIMLFAC method

1   If viewport specific layers are to be used, then:
   a) Make three new dimension layers using the same procedure as the exercise in this chapter, e.g. DIMTL, etc.
   b) Freeze in current viewports layers which have not to be 'displayed' using the same method as the exercise
   c) Making the appropriate viewport active then:
      1. restore the required UCS
      2. make the correct DIM layer current
      3. add the dimensions as shown – or your own

2   If paper space method to be used, then:
   a) model space dimension a linear part
   b) paper space dimension the same part
   c) divide the model value by the paper value to give DIMLFAC
   d) alter the DIMLFAC variable
   e) add the dimensions

3   Save the completed activities with a different name.

Chapter 26

Shading and 3D orbit

Shading and 3D orbit have been used in previous chapters without any discussion. In this chapter we will investigate both topics in greater detail.

Shading

Shading allows certain models to be displayed on the screen as a realistic coloured image. The models which can be shaded (and rendered) are 2½D extruded models, 3D objects, 3D surface models and solid models. We will use some previously created 3D surface models to investigate the topic.

1 Open the ruled surface model MODR2004\ARCHES from Chapter 16

2 Make the model tab active and restore UCS BASE. Any layer can be active

3 *a*) Menu bar with **View-Hide** to display the model with hidden line removal. Note the 2D icon is still displayed
  b) Menu bar with **View-Regen** to restore original model

4 *a*) Menu bar with **View-Shade-Hidden** and the model will be displayed with hidden line removal, but note the coloured 3D icon with X axis red, Y axis green and Z axis blue
  b) Menu bar with **View-Regen** and model is unchanged
  c) Menu bar with **View-Shade-3D Wire-frame** and the model will be displayed without hidden line removal, but the coloured 3D icon is still displayed
  d) Menu bar with **View-Shade-2D Wire-frame** to restore the original model with the 2D icon

5 Using the menu bar sequence View-Shade, activate the four shade options returning the display to 3D Wire-frame before activating the next option. Note the difference in the shading between the Flat and Gouraud options, evident at the arch curved 'shoulders'.

Explanation of the shade options

Activating SHADE from the View pull-down menu allows the user access to seven options. These options allow models to be displayed as shaded/wire-frame images as follows:

1 2D Wire-frame:
  The model is displayed with the boundaries as lines and curves with the 'normal' 2D icon. This option is generally used to restore shaded models to their original appearance.

2 3D Wire-frame:
  Models are displayed as lines and curves for their boundaries but with a coloured 3D icon. When used, this option restores shaded models to their original appearance but retains the coloured 3D icon.

3 Hidden:
   Displays models with hidden line removal and displays the coloured 3D icon. The REGEN command will not work when this option has been used, and models are restored to their original appearance with either the 2D Wire-frame or 3D Wire-frame options.

4 Flat Shaded:
   Models are shaded between their polygon mesh faces and appear flatter and less smooth than the Gouraud shaded models. Any materials (later chapter) which have been applied are also displayed flat shaded.

5 Gouraud Shaded:
   Models are shaded with the edges between the polygon mesh faces smoothed. This option gives models a realistic appearance. Added materials are also Gouraud shaded.

6 Flat Shaded, Edges On:
   Models are flat shaded with the wire-frame showing through the shade effect

7 Gouraud Shaded, Edges On:
   Models displayed with the Gouraud shading effect and the wire-frame shows through

8 General:
   a) Both the Flat and Gouraud shading options display the coloured 3D icon
   b) The Flat and Gouraud shading use the 2D wire-frame or the 3D Wire-frame options to restore the model to its original appearance
   c) The REDRAW/REGEN/REGENALL commands cannot be used if the View-Shade sequence is activated
   d) If HIDE <R> is entered from the command line to display any model with hidden line removal, then REGEN can be used

This completes the shading part of the chapter. Ensure that the ARCHES drawing is still displayed in the model tab for the 3D orbit exercise.

3D orbit

The 3D orbit command allows real-time 3D rotation, the user controlling this rotation with the pointing device. The basic concept is that you are viewing the model (the target) with a camera (the user), the target remaining stationary while the camera moves around the target. The opposite effect is displayed on the screen, i.e. the models appears to 'rotate'.

The 3D orbit concept has its own terminology, the most common being the arcball and icons – Fig. 26.1.

1 *Arcball*: is a circle with smaller circles at the quadrants

2 *Icons*: which alter in appearance dependent on were the pointing device is positioned relative to the arcball

3 *Click and drag*: the term for holding down the left button of the mouse and moving the mouse to give rotation of the model

4 *Roll*: a type of rotation when the icon is outside the arcball. It is a rotation about an axis through the centre of the arcball perpendicular to the screen.

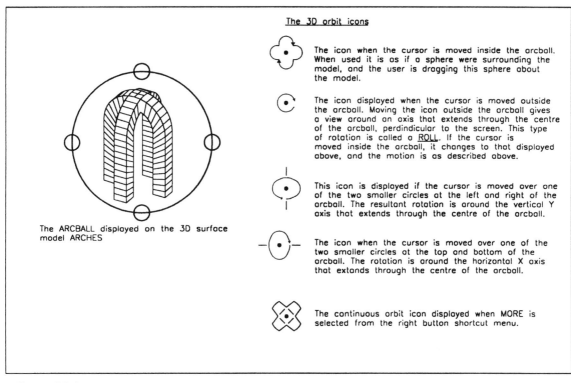

The 3D orbit icons

The icon when the cursor is moved inside the arcball. When used it is as if a sphere were surrounding the model, and the user is dragging this sphere about the model.

The icon displayed when the cursor is moved outside the arcball. Moving the icon outside the arcball gives a view around on axis that extends through the centre of the arcball, perdindicular to the screen. This type of rotation is called a ROLL. If the cursor is moved inside the arcball, it changes to that displayed above, and the motion is as described above.

This icon is displayed if the cursor is moved over one of the two smaller circles at the left and right of the arcball. The resultant rotation is around the vertical Y axis that extends through the centre of the arcball.

The icon when the cursor is moved over one of the two smaller circles at the top and bottom of the arcball. The rotation is around the horizontal X axis that extends through the centre of the arcball.

The continuous orbit icon displayed when MORE is selected from the right button shortcut menu.

The ARCBALL displayed on the 3D surface model ARCHES

Figure 26.1   The basic 3D orbit terminology.

Using 3D orbit

1   The ARCHES model should still be displayed in Model tab

2   Menu bar with **View-3D Orbit** and:
 prompt Arcball displayed
 respond a) hold down the left mouse button
    b) move the mouse about the screen
    c) release the mouse button
    d) practice this hold down, move, release and note the movement of the model
    e) press ESC to end command

3   Restore the model to its original viewpoint with U <R>

4   Menu bar with **View-Shade-Gouraud Shaded**

5   At the command line enter **3D ORBIT <R>** and:
 prompt Arcball displayed
 respond **right-click**
 prompt Shortcut menu
 respond **pick More**
 prompt cascade shortcut menu – Fig. 26.2
 respond **pick Continuous Orbit**
 prompt continuous icon displayed
 respond 1) move mouse slightly then leave it alone
    2) model displayed shaded with continuous rotation
    3) use the mouse left button to alter the rotation
    4) practice the continuous rotation
    5) right-click and pick Exit

6   Restore the model to the original orientation with U <R>

The 3D orbit shortcut menu

1   When the 3D orbit command is active, a right-click will display the shortcut menu which has been used to select the More and Exit options. The shortcut menu allows the user access to several useful and powerful options, the complete list being:

    a) Exit: does what it says, it exits the command

    b) Pan: the real-time AutoCAD pan command

    c) Zoom: the real-time AutoCAD zoom command, i.e. movement upwards gives magnification of model, movement downwards gives a reduction

    d) Orbit: indicates that the command is active/inactive

    e) More: allows the user access to an additional 10 options which include Adjust Distance, Swivel Camera, Continuous Orbit, Clipping Planes

    f) Projection: allows parallel or perspective selections

    g) Shading Modes: allows selection of the normal AutoCAD shading

    h) Visual Aids: compass, grid and UCS icon

    i) Reset View: a useful selection as it restores the model to its original orientation prior to using the orbit command and should be used instead of the U <R> used previously

    j) Preset Views: allows the normal 3D views to be activated

2   Using the shortcut menu is straightforward so:
    1. Activate the 3D orbit command and start rotating the model
    2. Press the right button
    3. Select the required option, e.g. Shading Modes-Flat Shaded
    4. Display restored and the 3D rotation can continue

3   The visual aids selection allows the user to display:
    a) a compass: a sphere is displayed inside the arcball consisting of three circles representing the X, Y and Z axes. This sphere rotates with the model.
    b) A grid: draws an array of lines on a plane parallel to the current X and Y axes, perpendicular to the Z axis.

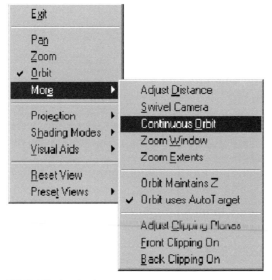

Figure 26.2   The 3D Orbit shortcut menu.

The clipping planes

The clipping plane options in 3D orbit are similar to the clip option of the DVIEW command, i.e. the clipping plane is an invisible plane set by the user. Parts of the model can be clipped, i.e. 'cut-away' relative to this clipping plane. To demonstrate the option:

1 Restore the model to its original orientation

2 Gouraud shade and start the model rotating with the 3D orbit command

3 Activate the shortcut menu and select the Hidden shading mode

4 Activate the shortcut menu and select More-Adjust Clipping Plane and:
   prompt   Adjust Clipping Plane 'dialogue box'
   respond 1) pick the Adjust Front Clipping icon
              2) drag the datum plane downwards – Fig. 26.3(a)
              3) right-click and close

5 Continue with the 3D orbit and the model will be rotated with a front clip effect – Fig. 26.3(b)

6 When you are satisfied with the display:
   a) right-click and reset view
   b) right-click and exit

7 The model should be restored to its original orientation

8 Try this clipping plane option a few times. It is relatively easy to use.

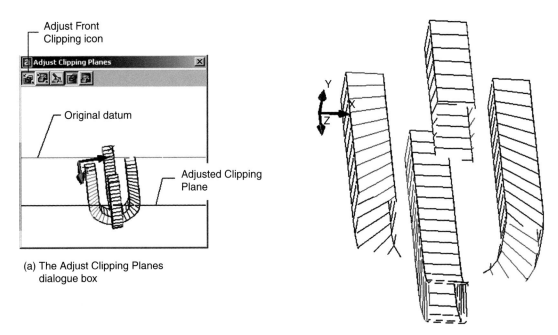

(a) The Adjust Clipping Planes dialogue box

(b) The ARCHES model after the front clip effect.

Figure 26.3   Using the 3D orbit option: More – Adjust Clipping Planes.

The 3D Orbit toolbar

The toolbar for the 3D orbit command is displayed in Fig. 26.4 and allows the user access to:

a) 3D pan and 3D zoom
b) the 3D orbit, swivel and continuous rotations
c) the 3D adjust distance
d) the clipping plane options: adjust, front and back
e) selecting the current 3D view

This completes the introduction to shading and using the 3D orbit command.

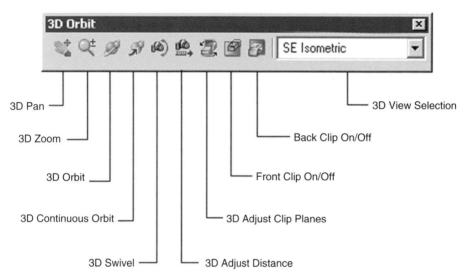

Figure 26.4   The 3D Orbit toolbar.

Summary

1 Surface and solid models can be shaded, the two options being Flat and Gouraud

2 Gouraud shaded models appear smoother than flat shaded models and this is the suggested mode for all future shading

3 The 3D orbit command allows interactive real-time 3D rotation of models – shaded or unshaded and with/without hidden line removal

4 The user has access to several useful options when the 3D orbit command is being used. These include parallel and perspective views of the model, front and back clipping and continuous 3D orbit.

5 The user should become familiar with using the 3D orbit command as it will be used with solid modelling

6 I would suggest that continuous orbit will be the most common option used with the 3D orbit command.

Chapter 27

Introduction to solid modelling

Three dimensional modelling with computer-aided draughting and design (CADD) can be considered in three categories, these being:

1 wire-frame

2 surface

3 solid

We have already created wire-frame and surface models and will now concentrate on how solid models are created.

This chapter will summarise the three model types.

Wire-frame modelling

1 Wire-frame models are defined by points and lines and are the simplest possible representation of a 3D component. They may be adequate for certain 3D model representation and require less memory than the other two 3D model types, but wire-frame models have several limitations:

   a) *Ambiguity*: it is difficult to know how a wire-frame model is being viewed, i.e. from above or from below?

   b) *No curved surfaces*: while curves can be added to a wire-frame model as arcs or trimmed circles, an actual curved surface cannot. Lines may be added to give a 'curved effect' but the computer does not recognise these as being part of the model.

   c) *No interference*: as wire-frame models have no surfaces, they cannot detect interference between adjacent components. This makes then unsuitable for kinematic displays, simulations, etc.

   d) *No physical properties*: mass, volume, centre of gravity, moments of inertia, etc. cannot be calculated.

   e) *No shading*: as there are no surfaces, a wire-frame model cannot be shaded or rendered.

   f) *No hidden line removal*: as there are no surfaces, it is not possible to display the model with hidden line removal.

2 AutoCAD 2004 allows wire-frame models to be created.

Surface modelling

1 A surface model is defined by points, lines and faces. A wire-frame model can be 'converted' into a surface model by adding these 'faces'. Surface models have several advantages when compared to wire-frame models, some of these being:

   a) Recognition and display of curved profiles

   b) Shading, rendering and hidden line removal are all possible, i.e. no ambiguity

   c) Recognition of holes

2 Surface models are suited to many applications but they have some limitations which include:
   a) *No physical properties*: other than surface area, a surface model does not allow the calculation of mass, volume, centre of gravity, moments of inertia, etc.
   b) *No detail*: a surface model does not allow section detail to be obtained.

3 Several types of surface model can be generated including:
   a) plane and curved swept surfaces
   b) swept area surfaces
   c) rotated or revolved surfaces
   d) splined curve surfaces
   e) nets or meshes

4 AutoCAD 2004 allows surface models of all these types to be created.

Solid modelling

1 A solid model is defined by the volume the component occupies and is thus a real 3D representation of the component. Solid modelling has many advantages which include:
   a) Complete physical properties of mass, volume, centre of gravity, moments of inertia, etc.
   b) Dynamic properties of momentum, angular momentum, radius of gyration, etc.
   c) Material properties of stress–strain
   d) Full shading, rendering and hidden detail removal
   e) Section views and profile extraction
   f) Interference between adjacent components can be highlighted
   g) Simulation for kinematics, robotics, etc.

2 Solid models are created using a **solid modeller** and there are several types of solid modeller, the two most common being:
   a) Constructive solid geometry or constructive representation, i.e. CSG/CREP. The model is created from solid primitives and/or swept surfaces using Boolean operations.
   b) Boundary representation (BREP). The model is represented by the edges and faces making up the surface, i.e. the topology of the component.

3 AutoCAD 2004 supports solid models of the CSG/CREP type.

4 The AutoCAD 2004 modeller is based on the **ACIS** solid modeller and supports **NURBS** – *non-uniform rational B splined curves*.

Comparison of the model types

The three model types are displayed in:
a) Figure 27.1: as models with hidden line removal
b) Figure 27.2: as model cross-sections

3D WIRE–FRAME MODEL	3D SURFACE MODEL	3D SOLID MODEL
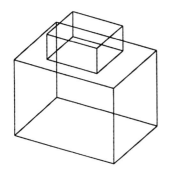	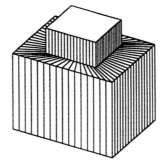	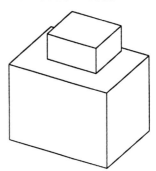
1. Model has length, width and height 2. There are no surfaces on the model. 3. The model does not have area, volume or mass 4. HIDE has no effect on the appearance of the model 5. The model displays AMBIGUITY ie it is difficult to know if the model is being viewed from above or from below.	1. Model has length, width and height 2. Surfaces have been added 3. The model has a surface area but no volume or mass 4. The HIDE command displays the model with hidden line removal 5. There is no ambiguity 6. No details can be extracted from the model. 7. The model can be shaded and rendered.	1. The model has length, width and height 2. The model has a surface area, a volume and a mass 3. The HIDE command will display the model with hidden line removal and there is no ambiguity 4. The model has mass properties eg centroid, moment of inertia, radius of gyration etc 5. Details can be extracted from the model eg sections 6. The model can be shaded and rendered

Figure 27.1   Simple comparison between wire-frame, surface and solid models.

3D WIRE–FRAME MODEL	3D SURFACE MODEL	3D SOLID MODEL
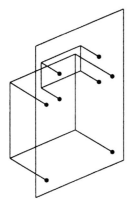	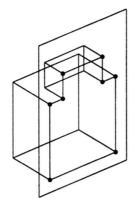	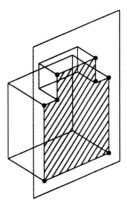
The model is represented as a series of corner points and edge lines. The cross–section of the model is a series of points which are not connected. The model is useful for general shapes, appearance and position only.	The model is represented as a series of corner points, edge lines and face surfaces. The cross–section of the model is a series of points and connected face lines. The model is useful for external surface display and can be displayed with hidden line removal and shading. Materials can be applied to the surfaces for rendering.	The model is represented as a series of corner points, edge lines, face surfaces and interior volume. The cross–section of the model is a series of points, face lines and section planes. The model is useful for mass properties, dynamic properties and material properties with hidden line removal and shading. Materials can be applied for rendering.

Figure 27.2   Further comparison of model types as cross-sections.

The solid model standard sheet

A solid model standard sheet (prototype drawing) will be created as a template and drawing file using the layouts from the surface model exercises, i.e. MV3DSTD. This standard sheet will:

a) be for A3 paper

b) have the four tab layouts: Model, MVLAY1, MVLAY2 and MVLAY3

1   Close any existing drawings or start AutoCAD

2   Open your MV3DSTD template or drawing file

3   Check the following:
   a) Tools-Named UCS: BASE, FRONT, RIGHT – set and saved
   b) Layers: 0,DIM,MODEL,OBJECTS,SECT,SHEET,TEXT,VP
   c) Sheet: layout to your own specification
   d) Text style: ST1 (romans.shx) and ST2 (Arial Black)
   e) Dimension style: 3DSTD with various settings

4   At the command line enter **-PURGE \<R>** and:
   prompt   Enter type of unused objects to purge
   enter   **LA \<R>** – layer option
   prompt   **Enter names to purge<*>** and **\<RETURN>**
   prompt   Verify each name to be purges [Yes/No] and enter: **Y \<R>**
   prompt   Purge layer 'DIM' and enter: **N \<R>**
   prompt   Purge layer 'OBJECTS' and enter: **Y \<R>**
   prompt   Purge layer 'SECT' and enter: **N \<R>**
   prompt   Purge layer 'TEXT' and enter: **N \<R>**

5   Repeat the command line -PURGE command with the following entries:
   a) B (blocks) and purge all (if any) blocks
   b) D (dimstyles) and purge all dimension styles except 3DSTD
   c) ST (text styles) and purge any text styles except ST1 and ST2
   d) SH (shapes) and purge all (if any) shapes
   e) M (multilines) and purge all (if any) multilines
   f) Note: entering PURGE at the command line will allow the user access to the Purge dialogue box. The various unwanted items can then be purged from the standard sheet.

6   At the command line enter **ISOLINES \<R>** and:
   prompt   Enter new value for ISOLINES<4>
   enter   **12 \<R>**

7   At the command line enter **FACETRES \<R>** and:
   prompt   Enter new value for FACETRES<0.5000>
   enter   **1 \<R>**

8   Display the Draw, Modify, Object Snap and Solids toolbars and position to suit

9   With the Model tab active, set a SE Isometric viewpoint

10   Make the MVLAY1 layout current and in model space with the lower left viewport active, restore UCS BASE with layer MODEL current

11 Menu bar with **File-Save As** and:
   prompt   Save Drawing As dialogue box
   respond   1. scroll at Files of type
               2. pick **AutoCAD Drawing Template File (*.dwt)**
   prompt   Save in AutoCAD Template folder dialogue box
   respond   1. enter file name as: **A3SOL.dwt**
               2. pick **Save**
   prompt   Template Description dialogue box
   respond   1. enter: **My solid model prototype created on ???**
               2. measurement: Metric
   then   pick OK

12 Menu bar with **File-Save As** and:
   prompt   Save Drawing As dialogue box
   respond   1. scroll at Files of type
               2. pick **AutoCAD 2004 Drawing (*.dwg)**
               3. scroll and pick your named folder
               4. enter file name: **A3SOL.dwg**
               5. pick **Save**

13 Menu bar with **File-Save As** and save you standard sheet as an AutoCAD Drawing Template file (A3SOL) in your named folder

14 You are now ready to start creating solid models.

Notes

1 The new A3SOL solid model standard sheet has been saved as both a template file and a drawing file

2 The A3SOL drawing file has been saved to your named folder, while the template file has been saved to both the AutoCAD Template folder and your named folder

3 Two new system variables have been introduced in the creation of the A3SOL template/drawing file, these being:
   ISOLINES: specifies the number of isolines per surface on objects. It is an integer with values between 0 and 2047. The default value is 4.
   FACETRES: adjusts the smoothness of shaded and rendered objects and objects with hidden line. The value can be between 0.01 and 10.00, the default being 0.5.

4 The ISOLINES and FACETRES values may be altered when creating some models

5 The viewports have not been zoom-centred, as this will depend on the model being created

6 Solid modelling consists of creating 'composites' from 'primitives' and AutoCAD 2004 supports the following types of primitive:
   a) basic   *b*) swept   *c*) edge

7 All three types of primitive will be investigated with examples and the various options for each will be discussed

8 While the A3SOL standard sheet has different layout configurations, I will generally only use the MVLAY1 or the MODEL tabs although the other layout tabs will be 'investigated' from time to time

Solid modelling is a fascinating topic, and should give the user a great deal of satisfaction as the various models are created and rendered.

Chapter 28

The basic solid primitives

AutoCAD 2004 supports the six basic solid primitives of box, wedge, cylinder, cone, sphere and torus. In this chapter we will create 'interesting' layouts using the six primitives. We will also investigate the various options which are available.

Note:
1 During the exercises do not just accept the co-ordinate values given. Try and reason out why they are being used

2 The procedure for each exercise is the same:
   a) open your A3SOL template file with layer MODEL and UCS BASE current
   b) ensure the Object Snap and Solid toolbars are displayed
   c) work (originally) with the Model tab
   d) save each completed exercise

The BOX primitive – Fig. 28.1

1 Select the BOX icon from the Solids toolbar and:
   prompt Specify corner of box or [CEnter]<0,0,0>
   enter   **0,0,0 <R>**
   prompt Specify corner or [Cube/Length]
   enter   **C <R>** – the cube option
   prompt Specify length
   enter   **100 <R>**
   and   a red cube will be displayed

2 Model tab should be active, so pan the cube to the lower centre of the screen

3 Select from the menu bar **Draw-Solids-Box** and:
   prompt Specify corner of box or [CEnter]<0,0,0>
   enter   **100,0,0 <R>**
   prompt Specify corner or [Cube/Length]
   enter   **L <R>** – the length option
   prompt Specify length and enter: **40 <R>**
   prompt Specify width and enter: **80 <R>**
   prompt Specify height and enter: **30 <R>**
   and   another red cuboid is displayed in all viewports

4 At the command line enter **CHANGE <R>** and:
   prompt Select objects
   respond **pick the smaller box then right click**
   prompt Specify change point or [Properties] and enter: **P <R>**
   prompt Enter property to change and enter: **C <R>**
   prompt Enter new color and enter: **3 <R> or green <R>**
   prompt Enter property to change and right-click/enter

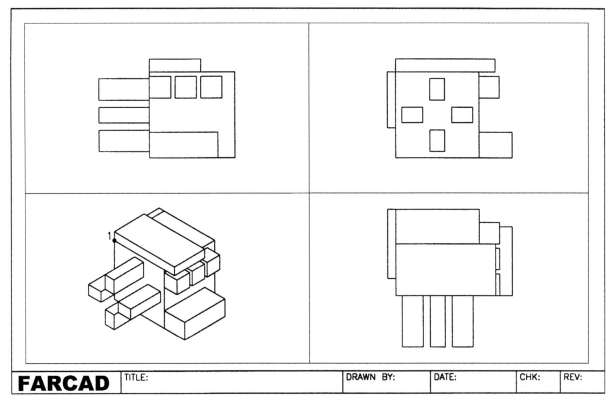

Figure 28.1 The BOX Primitive layout (BOXPRIM).

5 At the command line enter **BOX <R>** and:
   prompt   Specify corner of box or [CEnter]<0,0,0>
   respond **Endpoint icon and pick pt1** – first corner point
   prompt   Specify corner or [Cube/Length]
   enter   @120,60,15 **<R>** – the diagonally opposite corner point

6 Change the colour of this box to blue, with CHANGE <R> at the command line as step 4

7 Restore UCS RIGHT

8 Create a solid box with the following information:
   a) corner: 0,70,100
   b) cube option with length: 25

9 *a)* At the command line enter **CHANGE <R>** and:
     prompt   Select objects
     respond pick the last box created
     prompt   **1 was not parallel to the UCS**
     respond right-click to end the command
   b) At the command line enter **CHPROP <R>** and:
     prompt   Select objects
     respond pick the last box created and right-click
     prompt   Enter property to change and enter: **C <R>**
     prompt   New color [Truecolor/Colourbook] and enter: **MAGENTA <R>**
     prompt   Enter property to change and right-click/enter
   c) *Note*: The CHANGE and CHPROP commands will be discussed later. In future, when an object has to have its colour changed use the CHPROP command from the command line.

10 Select the Array icon from the Modify toolbar and:
   a) type of array: rectangular
   b) objects: the magenta box
   a) rows: 1 and columns: 3
   b) row offset: 0
   c) column offset: 35
   d) angle of array: −10

11 Restore UCS FRONT

12 Activate the solid BOX command and:
   prompt Specify corner of box or [CEntre]<0,0,0>
   enter   **CE <R>** – the box centre point option
   prompt Specify center of box and enter: **50,80,30 <R>**
   prompt Specify corner of box or [Cube/Length] and enter: **L <R>**
   prompt Specify length and enter: **18 <R>**
   prompt Specify width and enter: **25 <R>**
   prompt Specify height and enter: **60 <R>**

13 With CHPROP, change the colour of this last box to suit yourself. I used colour number 76

14 Polar array this last box with:
   a) method: items and angle to fill
   b) centre point: 50,50
   c) items: 4
   d) 360 angle with rotation

15 Finally restore UCS BASE and create another solid box with:
   a) corner: 0,100,100
   b) length: − 10; width: −80; height: −65
   c) colour: number 234

16 The model is now complete so:
   a) with MVLAY1 tab, zoom-extents then zoom to a factor of 1.5
   b) save the layout as **MODR2004\BOXPRIM**

17 *Investigate*:
   a) The model with hide – menu bar or command line
   b) Shading the viewports
   c) The 3D orbit command with the shaded model – model tab active
   d) Restoring the model to the original 3D wire-frame display

18 *Changing properties*
   When solid model layouts are being created the user will be asked to change some properties, especially the colour. All primitives will be originally displayed as red (due to layer MODEL being current) and thus require to be changed to another colour. This is to give a coloured image for shading, rendering and for using the 3D orbit command to maximum effect. Changing the colour of a primitive depends on the value of the PICKFIRST system variable and:
   a) if PICKFIRST is set to 0, then use CHANGE <R> at the command line as step 4 i.e. activate the command then pick the object which is to have its colour changed
   h) If PICKFIRST is set to 1 then:
      1 select the object to be changed
      2 pick the Properties icon from the Standard toolbar
      3 pick the Colour line
      4 scroll and pick the required colour
      5 cancel the Properties dialogue box and press ESC

19 *CHANGE or CHPROP?*
These two commands are similar but:
  a) CHANGE can only be used for objects created with the current UCS, hence the step 9(a) message
  b) CHPROP can be used with **ANY** UCS setting, and should be used for all future change colour operations

20 The user must now decide whether to set PICKFIRST to 0 or 1. My personal preference is to have PICKFIRST set to 0 and use the CHPROP command.

The WEDGE primitive – Fig. 28.2

1 Menu bar with **Draw-Solids-Wedge** and:
  prompt Specify corner of wedge or [CEnter]<0,0,0>
  enter   **0,0,0 <R>**
  prompt Specify corner or [Cube/Length]
  enter   **C <R>** – the cube option
  prompt Specify length
  enter   **100 <R>**
  and   red wedge displayed in all viewports

2 With the model tab, pan the wedge to lower centre of the screen

3 Select the WEDGE icon from the Solids toolbar and:
  prompt Specify corner of wedge or [CEnter]
  enter   **0,0,0 <R>**
  prompt Specify corner or [Cube/Length]
  enter   **@80, −60 <R>** – the other corner option
  prompt Specify height and enter: **50 <R>**

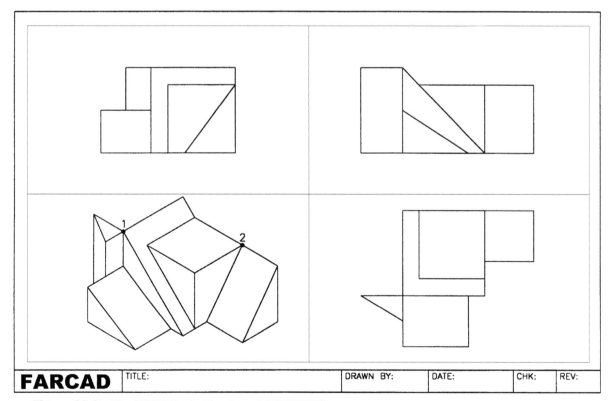

Figure 28.2 The WEDGE primitive layout (WEDPRIM).

4 Change the colour of this wedge to blue

5 At the command line enter **WEDGE <R>** and create a solid wedge with:
   a) corner: 100,100,0
   b) length: 60; width: 60; height: 80
   c) colour: green
   d) 2D rotate the green wedge about the point 100,100 by **−90degs**

6 Restore UCS FRONT and create a wedge with:
   a) corner: **endpoint icon and pick pt1**
   b) length: −50
   c) width: −100
   d) height: 30
   e) colour: magenta (CHPROP command)

7 Restore UCS BASE

8 The final wedge is to be created with:
   a) corner: **endpoint of pt2**
   b) cube option with length: −80
   c) colour: number 14

9 a) with MVLAY1 tab, zoom-extents then zoom 1.5
   b) save the model layout as **MODR2004\WEDPRIM**

10 Hide and shade the model in each viewport

11 Use 3D orbit with a shaded model in the model tab.

The CYLINDER primitive – Fig. 28.3

1 Menu bar with **Draw-Solids-Cylinder** and:
   prompt Specify center point for base of cylinder or [Elliptical]
   enter **0,0,0 <R>** – the cylinder base centre point
   prompt Specify radius for base of cylinder or [Diameter]
   enter **60 <R>** – the cylinder radius
   prompt Specify height of cylinder or [Center of other end]
   enter **100 <R>** – the cylinder height
   and a red cylinder is displayed, centred on 0,0,0

2 Pan to lower centre of screen

3 Select the CYLINDER icon from the Solids toolbar and:
   prompt Specify center point for base of cylinder or [Elliptical]
   enter **E <R>** – the elliptical option
   prompt Specify axis endpoint of ellipse for base of cylinder or
            [Center]
   enter **C <R>** – the centre option
   prompt Specify center point of ellipse for base of cylinder
   enter **80,0,0 <R>**
   prompt Specify axis endpoint of ellipse for base of cylinder
   enter **@20,0,0 <R>**
   prompt Specify length of other axis for base of cylinder
   enter **@0,30 <R>**
   prompt Specify height of cylinder or [Center of other end]
   enter **50 <R>**

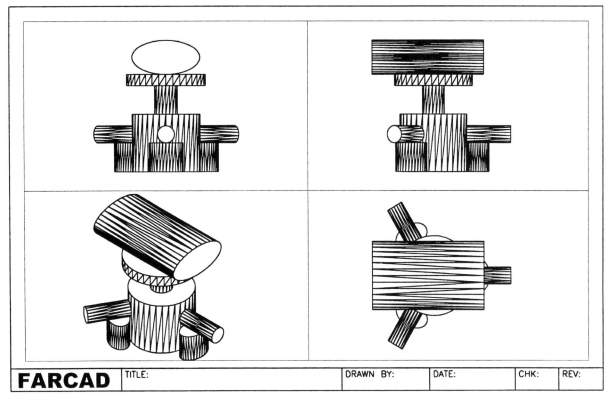

Figure 28.3 The CYLINDER primitive layout (CYLPRIM).

4 Change the colour of this cylinder to green, then polar array it:
   a) about the point 0,0
   b) items: 3
   c) angle: 360 with rotation

5 At the command line enter **CYLINDER <R>** and:
   prompt Specify center point for base of cylinder or [Elliptical]
   enter   **60,0,65 <R>** – the cylinder base centre point
   prompt Specify radius for base of cylinder and enter: **15 <R>**
   prompt Specify height of cylinder or [Center of other end]
   enter   **C <R>** – centre of other end option (*)
   prompt Specify center of other end of cylinder
   enter   **@80,0,0 <R>**

6 Change the colour of this cylinder to magenta and polar array it with the same entries
   as step 4

7 Create another two cylinders:
   centre pt   *rad*   *ht*   *colour*
   0,0,100   20   50   number 15
   0,0,150   70   20   blue

8 Finally create an **elliptical** cylinder:
   a) centre: 90,0,200
   b) axis endpoint: @60,0
   c) length of other axis: @0,30
   d) centre of other end: @ −200,0,0
   e) colour to suit

9 *a*) MVLAY1 tab and zoom-extents then zoom to a factor of 1
   b) save the model as **MODR2004\CYLPRIM**

10 Hide the model and note the triangular facets which are not displayed when the cylinder primitives are created. Shade then return the model to wire-frame representation

11 *Investigate*:
   a) With the 3D viewport active, enter ISOLINES at the command line and enter a value of 6. enter REGEN <R> and note model display
   b) change the ISOLINES value to 48 and regen
   c) return the ISOLINES value to the original 12
   d) enter FACETRES at the command line and alter the value to 2, then hide the model. Note the effect, then regen
   e) alter FACETRES to 5, hide, then regen
   f) return FACETRES to 1

12 *Note*:
   The ISOLINES system variable controls the appearance of primitive curved surfaces when they are created. The triangulation effect, or *FACETS*, is controlled by the system variable FACETRES. The higher the value of FACETRES (max 10) then the 'better the appearance' of curved surfaces, but the longer it takes for hide and shade. At our level, the values of 12 for ISOLINES and 1 for FACETRES are sufficient.

The CONE primitive – Fig. 28.4

1 Menu bar with **Draw-Solids-Cone** and:
   prompt Specify center point for base of cone or [Elliptical]
   enter   **0,0,0 <R>** – the cone base centre point
   prompt Specify radius for base of cone or [Diameter]
   enter   **50<R>**
   prompt Specify height of cone or [Apex]
   enter   **60 <R>**

2 Create another cone with:
   a) centre: 0,0,0
   b) radius: 90
   c) height: −80
   d) colour: green

3 Pan the model to the centre of the screen

4 Select the CONE icon from the Solids toolbar and:
   prompt Specify center point for base of cone or [Elliptical]
   enter   **E <R>** – the elliptical option
   prompt Specify axis endpoint of ellipse for base of cone or [Center]
   enter   **C <R>** – the centre point option
   prompt Specify center point of ellipse for base of cone
   enter   **70,0,0 <R>**
   prompt Specify axis endpoint of ellipse for base of cone and enter: **@20,0,0 <R>**
   prompt Specify length of other axis for base of cone and enter: **@0,25,0 <R>**
   prompt Specify height of cone or [Apex]
   enter   **35 <R>**

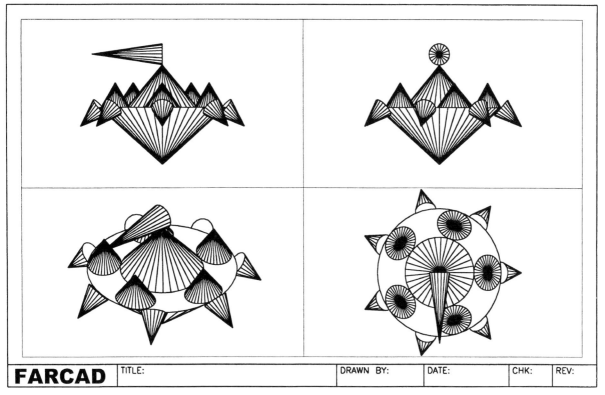

| **FARCAD** | TITLE: | | DRAWN BY: | DATE: | CHK: | REV: |

Figure 28.4   The CONE primitive layout (CONPRIM).

5 Change the colour of this cone to blue, then polar array it with:
  a) centre point: 0,0
  b) items: 5
  c) full angle to fill, with rotate items as copied active

6 At the command line enter **CONE <R>** and:
  prompt   Specify center point for base of cone and enter: **90,0,0 <R>**
  prompt   Specify radius for base of cone and enter: **15 <R>**
  prompt   Specify height of cone or [Apex]
  enter   **A <R>** – the apex option
  prompt   Specify apex point
  enter   **@40,0,0 <R>**

7 Change the colour of this last cone to colour number 235

8 Use the ROTATE 3D command with the last cone and:
  a) enter a Y axis rotation
  b) enter 90,0,0 as the point on the Y axis
  c) enter 41.63 as the rotation angle – *why this figure?*

9 Polar array the last cone about the point 0,0 for 7 items, full angle with rotate items active

10 Create the final cone with:
  a) centre: 0,0,75
  b) radius: 15
  c) apex option: @0,−100,0
  d) colour to suit

11 With MVLAY1 tab active, zoom-extents then zoom to a factor of 1.25 in all viewports

12 Hide, shade, 3D orbit then save as **MODR2004\CONPRIM**.

The SPHERE primitive – Fig. 28.5

1 Menu bar with **Draw-Solids-Sphere** and:
   prompt Specify center of sphere<0,0,0>
   enter **0,0,0 <R>**
   prompt Specify radius of sphere or [Diameter]
   enter **60 <R>**

2 Select the SPHERE icon from the Solids toolbar and:
   prompt Specify center of sphere<0,0,0>
   enter **80,0,0 <R>**
   prompt Specify radius of sphere or [Diameter]
   enter **D <R>** – the diameter option
   prompt Specify diameter
   enter **40 <R>**

3 Change the colour of this sphere to green, then polar array it:
   a) about the point: 0,0
   b) for 5 items
   c) full angle with rotate items as copied active

4 Pan the model to the centre of the screen

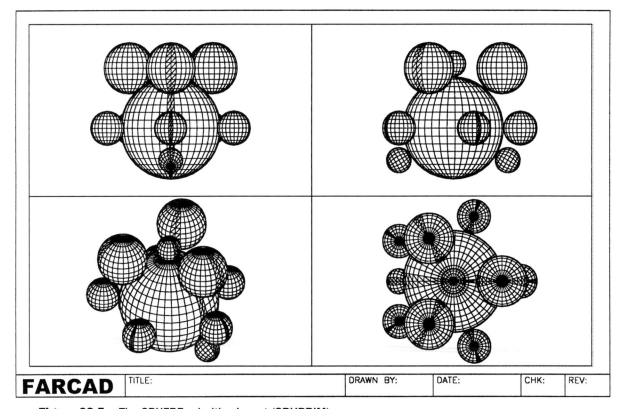

FARCAD	TITLE:		DRAWN BY:	DATE:	CHK:	REV:

Figure 28.5　The SPHERE primitive layout (SPHPRIM).

5 At the command line enter **SPHERE <R>** and create a sphere with:
   a) centre: 0,0,75
   b) radius: 15
   c) colour: magenta

6 Restore UCS FRONT and polar array the magenta sphere:
   a) centre point: 0,0
   b) items: 3
   c) full angle with rotate items as copied

7 Restore UCS BASE and create the final sphere with:
   a) centre: 58,0,70
   b) radius: 30
   c) colour: blue
   d) polar array about 0,0 for 3 items, full angle rotation

8 With the MVLAY1 tab, zoom-extents then zoom a factor of 1.5

9 Hide, shade, save as **MODR2004\SPHPRIM**

10 *Investigate*:
   Make any viewport (or the model tab) active and alter the following system variables, noting the sphere primitive appearance:
   a) ISOLINES set to 48 then regen
   b) ISOLINES set to 5 then regen
   c) Restore original ISOLINES 12
   d) FACETRES set to 10 then hide and regen.

The TORUS primitive – Fig. 28.6

1 Menu bar with **Draw-Solids-Torus** and:
   prompt Specify center of torus<0,0,0>
   enter   **0,0,0 <R>**
   prompt Specify radius of torus or [Diameter]
   enter   **80 <R>**
   prompt Specify radius of tube or [Diameter]
   enter   **15 <R>**

2 Restore UCS FRONT then select the TORUS icon from the Solids toolbar and:
   prompt Specify center of torus<0,0,0>
   enter   **80,0,0 <R>**
   prompt Specify radius of torus and enter: **50 <R>**
   prompt Specify radius of tube and enter: **20 <R>**

3 Change the colour of this torus to blue

4 Restore UCS BASE and polar array the blue torus about 0,0 for 3 items with full circle rotation and items copied

5 Restore UCS RIGHT, enter **TORUS <R>** at the command line and create a torus with:
   a) centre: 0,0,−95
   b) radius of torus: 80
   c) radius of tube: 20
   d) colour: green

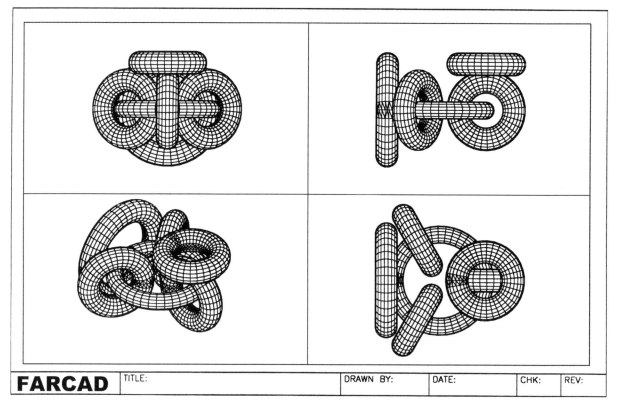

FARCAD	TITLE:		DRAWN BY:	DATE:	CHK:	REV:

Figure 28.6   The TORUS primitive layout (TORPRIM).

6 Restore UCS BASE and create the final torus with:
   a) centre: 80,0,80
   b) radius of torus: 50
   c) radius of tube: 20
   d) colour: to suit

7 MVLAY1 tab and zoom-extents then zoom to a factor of 1

8 Save as **MODR2004\TORPRIM**

9 *a*) With the lower left viewport active, HIDE and note the effect then regen
   b) Enter **DISPSILH <R>** at the command line and:
   prompt   Enter new value for DISPSILH <0>
   enter   **1 <R>**
   c) Hide the model, note the effect and compare this to that obtained in step (a)
   d) Now restore DISPSILH back to 0

10 *Note*; DISPSILH is a system variable which controls the display of silhouette curves of solid objects in wire-frame mode and:
   DISPSILH 0:   default value, models displayed 'as normal'
   DISPSILH 1:   models displayed with silhouette effect

11 It is your decision as to whether to have DISPSILH set to 0 or 1. I always have it set to 0.

Summary

1   The six solid primitives can be activated:
    a) from the menu bar with Draw-Solids
    b) by icon selection from the Solids toolbar
    c) by entering the solid name at the command line

2   The corner/centre start points can be:
    a) entered as co-ordinates from the keyboard
    b) referenced to existing objects
    c) picking suitable points on the screen

3   The six primitives have various options:
    box: *a*) corner; centre
               b) cube; length, width, height; other corner
    wedge: *a*) corner; centre
               b) cube; length, width, height; other corner
    cylinder: *a*) circular; elliptical
               b) radius; diameter
               c) height; centre of other end
    cone: *a*) circular; elliptical
               b) radius; diameter
               c) height; apex point
    sphere: *a*) centre only
               b) diameter; radius
    torus: *a*) centre only
               b) radius of torus; diameter
               c) radius of tube; diameter.

Assignment

An activity for your imaginative mind.

Activity 17: Using the six basic solid primitives.

You have to now create a layout **of your own design** with the six basic primitives using only the following information and with your own colour scheme:

Box	*Cylinder*	*Sphere*
length: 100	radius: 20	radius: 40
width: 40	height: 80	
height: 30		

Wedge	*Cone*	*Torus*
length: 50	radius: 50	torus radius: 50
width: 60	height: 120	tube radius: 20
height: 50		

I have shown two possibilities in this activity:
a) the traditional layout creation
b) an attempt at creating an 'everyday' object

Can you do better with the information given?

Chapter 29

The swept solid primitives

Solid models can be generated by extruding or revolving 'shapes' and in this chapter we will use several exercises to demonstrate how complex solids can be created from relatively simple shapes.

Extruded solids

Solid models can be created by extruding **CLOSED OBJECTS** such as polyline shapes, polygons, circles, ellipses, splines and regions:
a) to a specified height and taper angle
b) along a path

As with the solid primitive examples, each exercise should be started by opening the A3SOL template file with layer MODEL and UCS BASE current. The Solids (and other relevant) toolbars should be displayed. Work with the Model tab active then re-centre the MVLAY1 multi-viewport layout. Each exercise should be saved as a drawing file on completion.

Extruded Example 1: letters

1 With the Model tab active, change the viewpoint with the command line entry **VPOINT <R>** and:
    prompt Specify a view point or [Rotate]
    enter   **R <R>** – the rotate option
    then   enter angles of 30 and 30

2 Restore UCS RIGHT

3 Using the reference sizes given in Fig. 29.1:
    a) draw the three letters M, T and C as closed shapes using line and arc segments
    b) use **Modify-Object-Polyline** from the menu bar to 'convert' the three outlines into a single polyline with the Join option
    c) use your discretion for sizes not given (a snap of 5 helps)
    d) use the start points A, B and C given

4 Pan the three letters to the centre of the screen

5 Menu bar with **Draw-Solids-Extrude** and:
    prompt   Select objects
    respond **pick the letter M then right-click**
    prompt   Specify height of extrusion or [Path]
    enter   **80 <R>**
    prompt   Specify angle of taper for extrusion
    enter   **0 <R>**

6 The letter M will be extruded for a height of 80 in the positive Z direction

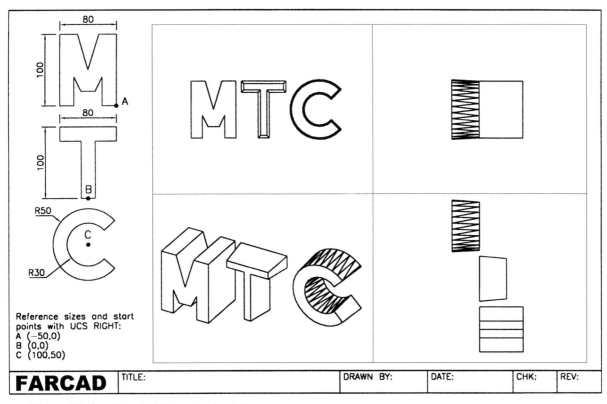

Figure 29.1 Extruded Example 1 – letters.

7 Select the EXTRUDE icon from the Solids toolbar and:
   prompt Select objects
   respond **pick the letter T then right-click**
   prompt Specify height of extrusion or [Path]
   enter **50 <R>**
   prompt Specify angle of taper for extrusion
   enter **5 <R>**

8 At the command line enter **EXTRUDE <R>** and:
   a) objects: pick the letter C then right-click
   b) height: enter **−50**
   c) taper angle: enter **−3**

9 Hide and shade with the model tab active

10 With MVLAY1 tab active:
   a) change the viewpoint in the lower left viewport as step 1
   b) zoom-extents then zoom to a factor of 1.1 in all viewports
   c) save the completed exercise using your own file name.

Extruded Example 2: keyed splined shaft

1 With model tab active and UCS BASE, refer to Fig. 29.2 and create two profiles:
   a) an outer tooth profile from two circles and an arrayed line, then trim as required.
      The circle centres should be at 0,0
   b) an inner shaft profile to your own specification

2 Use the menu bar sequence Modify-Object-Polyline (or PEDIT at the command line)
   to convert each profile into a single polyline using the Join option

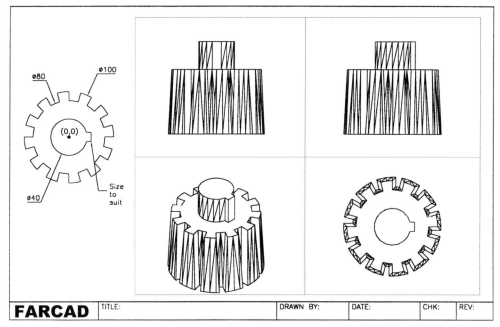

FARCAD | TITLE: | | DRAWN BY: | DATE: | CHK: | REV:

Figure 29.2 Extruded Example 2 – splined shaft.

3 Alter the system variables ISOLINES to 6 and FACETRES to 0.5

4 Select the EXTRUDE icon from the Solids toolbar and:
   prompt Select objects
   respond **pick the outer tooth profile then right-click**
   prompt Specify height of extrusion or [Path]
   enter **−70 <R>**
   prompt Specify angle of taper for extrusion
   enter **−3 <R>**

5 Repeat the Extrude icon selection and:
   a) objects: pick the inner shaft profile then right-click
   b) height: enter 30
   c) taper: enter 0

6 Note the 'denseness' of the splined model sides

7 Gouraud shade the model and use 3D orbit

8 Restore the original display with the menu bar sequence **View-Shade-2D Wire-frame**

9 MVLAY1 tab and zoom-extents then zoom to 2.5

10 Save the model layout to your named folder (own file name).

Extruded Example 3: a moulding

1 With the model tab active and UCS BASE, draw a polyline with:
   Start point: 0,0
   Next point: @0,100
   arc option with endpoint: @100,0
   arc endpoint: @100,−100
   line option with endpoint: @100,0 then right-click/enter

2 Change the colour of the polyline to blue and pan to suit

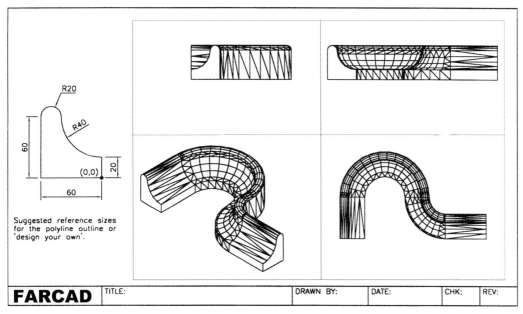

Figure 29.3 Extruded Example 3 – a moulding.

3 Restore UCS FRONT

4 Use the reference sizes in Fig. 29.3 to create the moulding as a single polyline or create your own outline design if required

5 Set ISOLINES to 12

6 Select the EXTRUDE icon from the Solids toolbar and:
   prompt   `Select objects`
   respond   **pick the red polyline then right-click**
   prompt   `Specify height of extrusion or [Path]`
   enter   **P <R>** – the path option
   prompt   `Select extrusion path`
   respond   **pick the blue polyline**
   and   the red outline is extruded along the blue path

7 Shade, 3D orbit then restore the original wire-frame model

8 MVLAY1 tab active and zoom-extents then zoom to a factor of 1

9 Save the model layout.

Extruded Example 4: a piping arrangement

1 With the model tab, layer MODEL and UCS BASE, create an 8 segment 3D polyline with the menu bar sequence **Draw-3D Polyline** and:
   prompt   `Specify start point of polyline` and enter: **0,0 <R>**
   prompt   `Specify endpoint of line` and enter: **@0,0,100 <R>**
   prompt   `Specify endpoint of line` and enter: **@100,0,50 <R>**
   prompt   `Specify endpoint of line` and enter: **@0,100,50 <R>**
   prompt   `Specify endpoint of line` and enter: **@ −100,0,50 <R>**
   prompt   `Specify endpoint of line` and enter: **@0,−75,50 <R>**
   prompt   `Specify endpoint of line` and enter: **@75,0,50 <R>**
   prompt   `Specify endpoint of line` and enter: **@0,50,50 <R>**
   prompt   `Specify endpoint of line` and enter: **@0,0,100 <R>**
   prompt   `Specify endpoint of line` and: **right-click/enter**

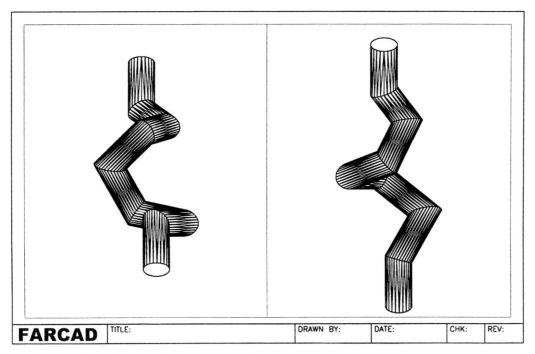

FARCAD	TITLE:		DRAWN BY:	DATE:	CHK:	REV:

Figure 29.4   Extruded Example 4 – a piping arrangement.

2   Change the colour of the 3D polyline to green

3   Draw a circle, centred on 0,0 with a radius of 25

4   Copy the circle and 3D polyline from 0,0 to 200,0

5   With the menu bar sequence **Modify-Object-Polyline**, spline the copied 3D polyline

6   Select the EXTRUDE icon and:
   a) objects: pick the first red circle then right-click
   b) enter P <R> for the path option
   c) path: pick the first green 3D polyline

6   The circle will be extruded along the green path to give a piping arrangement

7   *Question*: Green polyline but red model – why?

8   Repeat the extrude command and:
   a) select the copied red circle as the object
   b) enter P <R> for the path option
   c) pick the copied green polyline as the path
   d) no extruded model results and the command line displays the message:
      Cannot extrude along 3D spline path
      Cannot extrude along this path

9   Gouraud shade the model, 3D orbit then return to wire-frame

10   Select a layout tab and centre the model. Fig. 29.4 gives the MVLAY2 configuration for the piping arrangement, thus giving two views of the model, one from above and another from below.

11 When an object is extruded (or revolved) the original 'shape' is either retained in the drawing or deleted from the drawing. The effect is controlled by the system variable **DELOBJ** and:
DELOBJ: 0, shape is retained
DELOBJ: 1, shape is deleted

This completes the extrusion exercises.

Revolved solids

Solid models can be created by revolving objects (closed polylines, polygons, circles, ellipses, closed splines and regions):
a) about the X and Y axes
b) about selected objects by a specified angle

As with extrusions, very complex models can be obtained from relatively simple shapes.

Revolved Example 1: a flagon (of sorts)

1 A3SOL with model tab, UCS BASE and set ISOLINES to 8

2 Restore UCS FRONT and make and refer to Fig. 29.5

3 Draw a **closed** polyline outline using the sizes given or, ideally, design your own outline. Use the (0,0) start point indicated

4 Menu bar with **Draw-Solids-Revolve** and:
  prompt Select objects
  respond **pick the polyline then right-click**
  prompt Specify start point for axis of revolution or define axis
            by [Object/X (axis)/Y (axis)]
  enter   **Y <R>** – the Y axis option
  prompt Specify angle of revolution
  respond **right-click** to accept the 360 (full circle) default

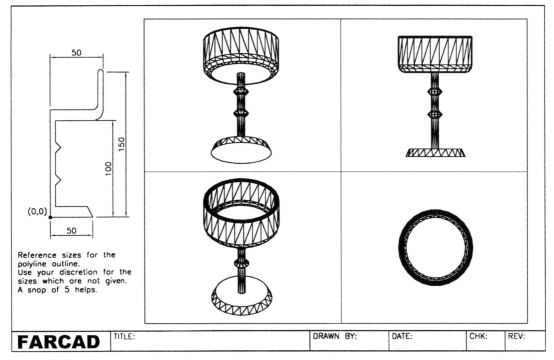

Figure 29.5 Revolved Example 1 – a flagon.

5 The polyline outline will be revolved into a swept revolved solid

6 Gouraud shade the model then use the 3D orbit command

7 With the MVLAY1 tab active, alter the viewpoint angle in the left viewport to view the model from below with the command line entry **VPOINT** then use the ROTATE option with entered angles of 30 and −30

8 In each viewport, zoom-extents then zoom to a factor of 1.5

9 Save the completed model entering your own file name.

Revolved Example 2: a shaft

1 A3 template file with usual settings – model tab, UCS BASE and layer model current

2 Decide on the system variable DELOBJ value – 0 or 1

3 Make layer TEXT (green) current and refer to Fig. 29.6

4 Draw a **closed** polyline outline using the sizes given. Use the (20,−30) start point. Note that I have only displayed the 3D viewport with this exercise. This was for convenience purposes only.

5 Set ISOLINES to 12 and make layer MODEL current

6 Menu bar with **Draw-Solids-Revolve** and:

prompt	Select objects
respond	**pick the polyline then right-click**
prompt	Specify start point of axis of revolution or define axis by [Object/X(axis)/Y(axis)]
enter	**X <R>** – the X axis option
prompt	Specify angle of revolution
enter	**360 <R>**

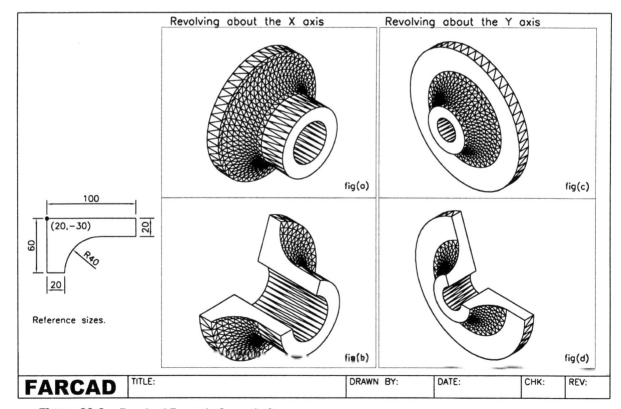

Figure 29.6 Revolved Example 2 – a shaft.

7 The polyline outline will be revolved about the X axis as Fig. 29.6(a)

8 Observe the effect then undo the revolved effect to restore the original green polyline outline

9 Select the revolve icon from the solids toolbar and:
   a) objects: select the polyline outline then right-click
   b) axis of revolution: X axis
   c) angle of revolution: 250
   d) the revolved effect will be as Fig. 29.6(b). Observe then undo

10 At the command line enter **REVOLVE <R>** and:
   a) objects: select polyline then right-click
   b) axis of revolution: enter Y <R>
   c) angle of revolution: 360
   d) the result will be as Fig. 29.6(c) – pan needed?
   e) observe then undo the revolved effect

11 Finally revolve the polyline:
   a) about the Y axis
   b) with a 250 angle of revolution – Fig. 29.6(d)

12 Shade, 3D orbit, etc. then centre the MVLAY1 tab configuration

13 Save the model if required

   This completes the swept primitive exercises.

Plotting multiple viewport layouts with hide

To obtain a plot of a multiple layout with hide, the user needs the use the **MVIEW** command. To demonstrate this command:

1 Open any multiple viewport drawing, e.g. the splined shaft example

2 Enter paper space

3 At the command line enter **MVIEW <R>** and:
   prompt   Specify corner of viewport or [various options]
   enter   **H <R>** – the hideplot option
   prompt   Hidden line removal for plotting [ON/OFF]
   enter   **ON <R>**
   prompt   Select objects
   respond **pick the *borders* of all viewports which have to have hidden line removal**
   then   right-click

4 The command line is returned

5 Proceed with the normal plot sequence, but ensure that the Hide Plot option is **NOT ACTIVE** (i.e. no tick)

6 This procedure should be used at all times when multi-viewport plots are required with hidden line removal.

Summary

1 Swept solids are obtained with the extrude and revolve commands

2 The two commands can be activated:
   a) by icon selection from the Solids toolbar
   b) from the command line with Draw-Solids
   c) by entering EXTRUDE and REVOLVE at the command line

3   Very complex models can be obtained from simple shapes

4   Only certain 'shapes' can be extruded/revolved. These are closed polylines, circles, ellipses, polygons, closed splines and regions (more on this in a later chapter).

5   Objects can be extruded:
    a) to a specified height
    b) with/without a taper angle
    c) along a path curve

6   The extruded height is in the Z direction and can be positive or negative

7   The taper angle can be positive or negative

8   Objects can be revolved:
    a) about the X and Y axes
    b) about an object
    c) by specifying two points on the axis of revolution

9   The angle of revolution can be full (360) or partial.

Assignment

During the excavation of the ancient city of CADOPOLIS, the intrepid diggers uncovered two artefacts, both of which they attributed to our master builder MACFARAMUS. They decided (how we will never know) that the artefacts were scale models of a section from a pillar and the partial wheel from a chariot. It is these that you have to create as solid models from swept primitives.

Activity 18: Two swept primitive models designed by MACFARAMUS.

These two models have to be created from closed polylines. The dimensions taken by the site engineers were not complete and only the basic sizes have been given. You have to use your own discretion when drawing the two outlines. The procedure for both models is:

1   Open your A3SOL template file as normal

2   Draw the outline of the model as a closed polyline, both with UCS BASE, layer MODEL and with the model tab active

3   Extrude the pillar outline for a height of 120 with a 5 taper

4   Revolve the wheel outline about the X or Y axis dependant on how the original outline was drawn. The partial angle of revolution is to be 4 PI radians.

5   Centre the models in the layout tabs

6   *Note*: I have displayed both models on the one sheet of paper with:
    a) pillar: 3D and top views
    b) wheel: two 3D views, from above and from below

7   When complete, save each model.

Chapter 30

Boolean operations and composite solids

The basic and swept solids which have been created are called **primitives** and are the 'basic tools' for solid modelling. With these primitives the user can create **composite solids**, so called because they are 'composed' of two or more solid primitives, i.e.

a) primitive: a box, wedge, cylinder, extrusion, etc.

b) composite: a solid made from two or more primitives.

Composite solids are created from primitives using the three **Boolean** operations of union, subtraction and intersection. Figure 30.1 demonstrates these operations with two primitives:

a) a box

b) a cylinder 'penetrating' the box.

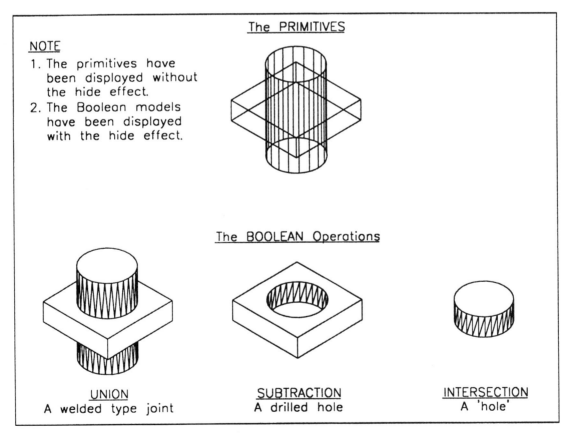

NOTE
1. The primitives have been displayed without the hide effect.
2. The Boolean models have been displayed with the hide effect.

The PRIMITIVES

The BOOLEAN Operations

UNION
A welded type joint

SUBTRACTION
A drilled hole

INTERSECTION
A 'hole'

Figure 30.1   The three BOOLEAN operations.

Union

1   This operation involves 'joining' two or more primitives to form a single composite, the user selecting *all objects to be unioned*

2   The operation can be considered similar to welding two or more components together.

Subtraction

1   This involves removing one or more solids from another solid thereby creating the composite. The user selects:
    a) *the source solid*
    b) *the solids to be subtracted from the source solid*

2   The result of a subtraction operation can be likened to a drilled hole, i.e. if the cylinder is subtracted from the box, a hole will be obtained in the box

3   *Note*: the source solid is generally 'the larger solid' as you cannot normally take a large solid from a small solid.

Intersection

1   This operation gives a composite solid from other solids which have a common volume, the user selecting *all objects which have to be intersected*

2   The box/cylinder illustration of the intersection operation gives a 'disc shape' or 'hole', i.e. if the box and cylinder are intersected, the common volume is the disc shape.

Creating a composite solid from primitives

There is no 'correct or ideal' method of creating a composite, i.e. the Boolean operations selected by one user may be different from those selected by another user, but the final composite may be the same. To demonstrate this, we will create an L-shaped component by three different methods, so:

1   Open your A3SOL template file refer to Fig. 30.2 and:
    a) activate the MVLAY1 tab
    b) enter paper space
    c) erase any text and the four viewports
    d) with layer VP current create a single viewport with:
        i)   first point: pick to suit in lower left corner area
        ii) other corner: enter @360,220
    e) return to model space, UCS BASE, layer MODEL
    f) set a SE Isometric viewpoint

2   Create two box primitives:
    a) corner: 0,0,0   *b*) corner: 100,100,100
        cube option   length: −60
        length: 100   width: −100
        colour: red   height: −70
                                    colour: blue

3   Create another two box primitives:
    a) corner: 125,125,0   *b*) corner: 125,125,30
        length: 100   length: 40
        width: 100   width: 100
        height: 30   height: 70
        colour: red   colour: blue

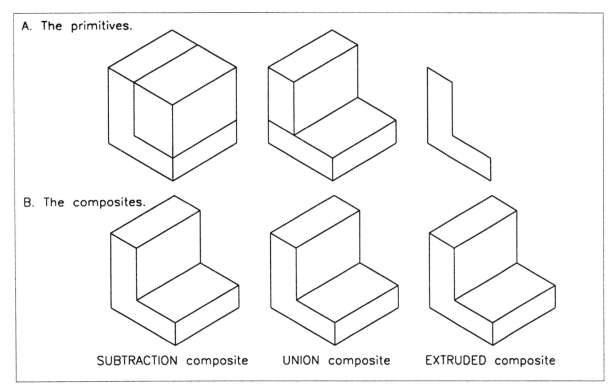

A. The primitives.

B. The composites.

SUBTRACTION composite   UNION composite   EXTRUDED composite

Figure 30.2 Creating the same composite by three different methods.

4 Restore UCS FRONT and draw a 2D polyline shape:
   Start point: 500,250   *Next point*: @100,0   *Next point*: @0,30
   Next point: @−60,0   *Next point*: @0,70   *Next point*: @−40,0
   Next point: close

5 Restore UCS BASE and centre the viewport with the zoom-centre option about the
   point 100,250,0 at 400 magnification.

6 From the menu bar select **Modify-Solids Editing-Subtract** and:
   prompt   Select solids and regions to subtract from...
            Select objects
   respond **pick the left red box then right-click**
   prompt   Select solids and regions to subtract...
            Select objects
   respond **pick the left blue box then right-click**
   and   the blue box is subtracted from the red box

7 Menu bar with **Modify-Solids Editing-Union** and:
   prompt   Select objects
   respond **pick the middle red and blue boxes then right-click**
   and   the two boxes will be unioned

8 Restore UCS FRONT and:
   a) select the EXTRUDE icon from the Solids toolbar
   b) pick the L-shaped polyline then right-click
   c) enter an extruded height of −100 with 0 taper
   d) the L-shaped polyline is extruded into a composite

9 *Task*

a) Hide the models – all the same?

b) Gouraud shade the models and note the different colour effect between the union and subtraction composites – any comment on this colour effect? Think order of pick!

c) Menu bar with **View-Shade-2D** Wire-frame to 'return' the model to its 2D wire-frame representation

d) Menu bar with **Tools-Inquiry-Region/Mass Properties** and:

prompt	Select objects
respond	**pick left composite then right-click**
and	AutoCAD Text Window with:
	Mass: 580000.00
	Volume: 580000.00
	Bounding box, Centroid, etc.
enter	**N <R>** in response to 'Write analysis to file' prompt

e) Repeat the MASSPROP command and select the middle and right composites – same mass and volume?

10 *Questions*

a) Why are the mass and volume the same?
Answer: AutoCAD 2004 assumes a density value of 1 and does not support different material densities

b) Is the volume of 580000 correct for the L shape?

c) What are the volume units?

11 Now that we have investigated the Boolean operations, we will create some composite solid models which (I hole) will be interesting.

Summary

1 There are three Boolean operations – union, subtraction and intersection

2 The three operations can be activated:
a) from the menu bar with Modify-Solids Editing
b) in icon form from the Solids Editing toolbar
c) by entering the command from the keyboard

3 The Boolean operations are derived from Boolean Algebra (set theory) and are essential for the AutoCAD creation of solid composites from primitives.

Chapter 31

Composite model 1 – a machine support

In this exercise we will create a composite solid from the box, wedge and cylinder primitives using the three Boolean operations. Once created, we will dimension the model using viewport specific layers.

The exercise is quite simple and you should have no difficulty in following the various steps in the model construction. Try and work out why the various entries are given – do not just accept them.

1 Open your A3SOL template file with normal settings and display the Solids, Solids Editing and other toolbars to suit

2 Refer to Fig. 31.1 which displays only the 3D viewport of the model at various stages of its construction

3 In this exercise we will work with the MVLAY1 tab active, so in each viewport, zoom-extents then zoom to 1.5 scale

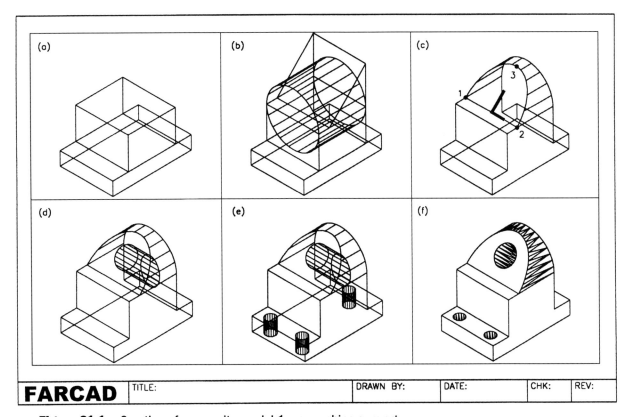

Figure 31.1   Creation of composite model 1 – a machine support.

4 With the 3D viewport active, layer MODEL and UCS BASE, use the BOX icon to create two primitives as Fig. 31.1(a) with:

   a) corner: 0,0,0   *b*) corner: 100,120,25
      length: 100   length: −100
      width: 150   width: −90
      height: 25   height: 60
      colour: red   colour: blue

5 Create a cylinder on top of the blue box with:

   a) centre: 50,30,85
   b) radius: 50
   c) height/other end: enter **C <R>** then **@0,90,0 <R>**
   d) colour: green

6 Create a wedge with:

   a) corner: 0,120,85
   b) length: 70; width: 100; height: 70
   c) colour: magenta

7 Rotate (2D rotate) the magenta wedge:

   a) about the point: 0,120,85
   b) by: −90 degrees to give Fig. 31.1(b)

8 Select the INTERSECTION icon from the Solids Editing toolbar and:

   prompt   Select objects
   respond **pick the green cylinder and magenta wedge then right-click**

9 Select the UNION icon from the Solids Editing toolbar and:

   prompt   Select objects
   respond **pick the red and blue boxes and the intersected wedge/cylinder then right-click**

10 The model now appears as Fig. 31.1(c)

11 Refer to Fig. 31.1(c) and menu bar with **Tools-New UCS-3 Point** and set a new UCS with:

   a) origin: MIDpoint icon and pick line 12
   b) X axis: ENDpoint icon and pick pt2
   c) Y axis: QUADrant icon and pick pt3 on curve

12 The UCS icon will move and align itself on the sloped surface
   Note: if icon does not move, menu bar with View-Display-UCS Icon and ensure that On and Origin are ticked

13 Save this UCS position as SLOPE

14 Create a cylinder with:

   a) centre: 0,35,0
   b) radius: 18
   c) height: −100
   d) colour: number 54 – use the CHPROP command

15 Select the SUBTRACT icon from the Solids Editing toolbar and:

   prompt   Select solids and regions to subtract from ..
             Select objects
   respond **pick the composite model then right-click**
   prompt   Select solids or regions to subtract
             Select objects
   respond **pick the cylinder then right-click** – Fig. 31.1(d)

16 Restore UCS BASE

17 Create a cylinder with:
 a) centre: 20,15,0
 b) radius: 9
 c) height: 25
 d) colour: number 174

18 Multiple copy the cylinder:
 a) from: 20,15
 b) by: @60,0 and by: @30,120

19 Using the SUBTRACT icon:
 a) select the original composite then right-click
 b) pick the three cylinders then right-click

20 The model is now complete and is displayed in Fig. 31.1(e) without hide and in Fig. 31.1(f) with hide

21 Select the Model tab and Gouraud shade the model, then use the 3D orbit command to rotate the shaded model

22 Restore the model to 2D wire-frame representation, then make the MVLAY1 tab active

23 At this stage save the composite as **MODR2004\MACHSUPP**.

Dimensioning the model

There are several methods which can be used to add dimensions to a multi-view solid model and in this chapter we will use viewport specific layers. Later chapters will introduce the user to the other methods.

The MVLAY1 tab displays the model in a four viewport configuration and we now want to add dimensions to three of these viewports. To create the viewport specific layers:

1 At the command line enter **VPLAYER <R>** and:
 prompt   Enter an option [?/Freeze/Thaw/Reset/Newfrz/Vpvisdflt]
 enter   **N <R>** – the Newfrz (new viewport freeze) option
 prompt   Enter name(s) of new layers frozen in all viewports
 enter   **DIMTL,DIMTR,DIMBR <R>**
 prompt   Enter an option [?/Freeze/Thaw/..
 respond   **<RETURN>** – no other option to be used

2 Make the top left viewport active, menu bar with **Format-Layer** and:
 prompt   Layer Properties Manager dialogue box
 note   three new layers – DIMBR, DIMTL, DIMTR with:
       i) Frozen in current viewport – blue icon
       ii) Frozen in new viewport – blue icon
       iii) You may have to stretch the dialogue box
 respond   1. activate Show Details
       2. **pick DIMTL** (highlights) and note details – Freeze in current viewport is on (tick)
       3. **pick blue icon Freeze in active viewport** to Thaw the layer and the tick is removed from Freeze in current viewport details list
       4. pick OK

3   With the top right viewport active, Format-Layer and:
    a) pick layer **DIMTR**
    b) remove the tick at Freeze in current viewport to Thaw the layer
    c) pick OK

4   With lower right viewport active, Format-Layer and:
    a) pick layer **DIMBR**
    b) toggle the current viewport Freeze to Thaw
    c) pick OK

5   What has been achieved in this section?
    a) three new viewport specific layers have been made
    b) these layers have been named DIMTL for the top left viewport, DIMTR for the top right viewport and DIMBR for the bottom left viewport
    c) the three layers were originally created:
       i) frozen in new viewports
       ii) currently frozen in all viewports
    d) each layer was currently thawed in a specific viewport, e.g. layer DIMTL is currently thawed in the top left viewport but is currently frozen in the other three viewports. Layers DIMTR and DIMBR are currently frozen in the top left viewport.

6   Before adding the dimensions, change the colour of the three new layers (DIMTL, DIMTR, DIMBR) to magenta using the Layer Properties Manager dialogue box.

Adding the dimensions

1   Before the dimensions are added to the model, menu bar with **Dimension-Style** and using the Dimension-Style Manager dialogue box:
    a) 3DSTD the current (and only) style?
    b) pick Modify
    c) pick the Fit tab
    d) Alter Scale for Features: Use overall scale of 2
    e) pick OK to return to Dimension-Style Manager dialogue box
    f) pick the 3DSTD style name then Set Current
    g) pick Close
    h) this will scale all the dimension parameters by 2

2   Make the lower right viewport active and:
    a) restore UCS BASE
    b) make layer DIMBR current
    c) refer to Fig. 31.2

3   With Dimension-Linear from the menu bar, or with the LINEAR dimension icon from the Dimensions toolbar, add:
    a) the horizontal dimension
    b) the four vertical dimensions using the baseline option

4   With the top right viewport active:
    a) restore UCS FRONT
    b) make layer DIMTR current
    c) add the two linear dimensions

5   Make the top left viewport active and:
    a) restore UCS RIGHT
    b) make layer DIMTL current
    c) add the six dimensions

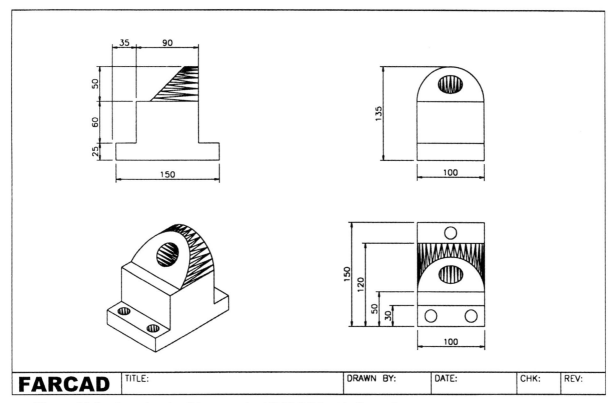

Figure 31.2 Completed solid composite MACHSUPP with dimensions.

6   The composite model is now complete with dimensions added and can be plotted with the layer VP frozen for effect – Fig. 31.2

7   These dimension additions do not need to be saved.

This completes the first composite model exercise.

Chapter 32

Composite model 2 – a backing plate

In this exercise we will create a solid from an extruded swept primitive and then subtract various 'holes' to complete the composite. The exercise will also involve altering the viewport layout of the A3SOL template file. As with all the exercises, do not just accept the entries – work out why the various values are being used.

The model

Refer to Fig. 32.1 which details the model to be created and gives the relevant sizes. As an exercise, draw the three orthogonal views as given and then add the isometric (the arc 'hole' is interesting to complete in an isometric view). Time how long it takes to complete this 2D drawing. I spent about an hour to complete the four views with dimensions.

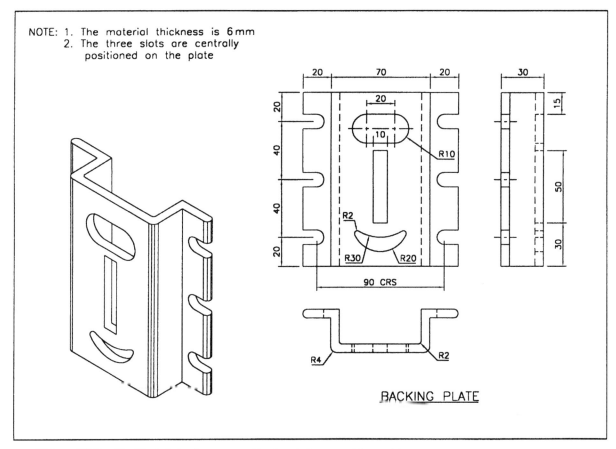

Figure 32.1   Backing plate drawn as orthogonal views and isometric.

Setting the viewports

1   Open your A3SOL template file and with MVLAY1 tab active:
   a)   enter paper space
   b)   erase the four viewports
   c)   make layer VP current
   d)   refer to Fig. 32.2

2   Menu bar with **View-Viewports-1 Viewport** and:
   prompt   Specify corner
   enter   **10,25 <R>**
   prompt   Specify opposite corner
   enter   **@155,175 <R>**

3   Create another three single viewports using the following co-ordinate entries:
   a)   first corner: 165,25
        opposite corner: @160,70
   b)   first corner: 165,95
        opposite corner: @160,155
   c)   first corner: 325,95
        opposite corner: @70,155

4   In model space:
   a)   make layer MODEL current
   b)   set UCSVP to 0 in each viewport
   c)   set the 3D viewpoints in the viewports as Fig. 32.2(a)
   d)   restore UCS BASE
   e)   zoom-centre about 0,10,60 at **1XP** – yes enter 1XP

5   Now activate the Model tab and pan the icon to centre of the screen.

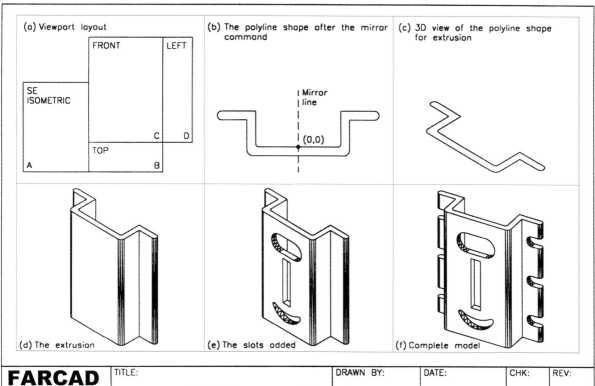

Figure 32.2   Steps in the creation of the backing plate composite.

Creating the extrusion

1   Using the polyline icon from the Draw toolbar, create a single polyline from line and
    arc segments with the following entries:
    Start point: 0,0
    Next point: @27,0
    Next point: Arc option, with endpoint of arc @2,2
    Next point: Line option, to @0,22
    Next point: @23,0
    Next point: Arc option, with endpoint of arc @0,−6
    Next point: Line option, to @−17,0
    Next point: @0,−20
    Next point: Arc option, with endpoint of arc @−4,−4
    Next point: Line option, to @ 31,0
    Next point: right-click/enter

2   2D mirror the polyline shape about the points 0,0 and 0,50 and do not delete source
    objects

3   Use the menu bar sequence **Modify-Object-Polyline** and:
    a) pick any point on the right-hand polyline
    b) enter **J <R>** – the join option
    c) pick the two polyline shapes then right-click
    d) enter **X <R>** to end the command

4   The two 'halves' of the polyline have been joined into a single polyline as Fig. 32.2(b)
    in plan view and Fig. 32.2(c) in 3D

5   At the command line enter **ISOLINES <R>** and:
    prompt   Enter new value for ISOLINES<24>
    enter   **3 <R>**

6   *Note*:
    the ISOLINES system variable has been reduced to 3 due to the 'corner edges' of the
    model. With the original value, the extrusion would result in these corner edges being
    'very densely' displayed

7   Select the EXTRUDE icon from the Solids toolbar and:
    prompt   Select objects
    respond   **pick any point on the polyline then right-click**
    prompt   Specify height of extrusion or [Path]
    enter   **120 <R>**
    prompt   Specify angle of taper for extrusion
    enter   **0 <R>**

8   The polyline shape will be extruded as Fig. 32.2(d)

9   With the model tab active, pan the model to suit.

Creating the 'holes'

1   Still with the Model tab active, restore UCS FRONT

2   Create a box primitive with:
    a) corner: −5,30,0
    b) length: 10; width: 50; height: 6
    c) colour: blue

3 Create the following three primitives:

box	*cylinder*	*cylinder*
corner: −10,85,0	centre: −10,95,0	centre: 10,95,0
length: 20	radius: 10	radius: 10
width: 20	height: 6	height: 6
height: 6	colour: green	colour: green
colour: green		

4 Union the three green primitives

5 Draw two circles:
   a) centre: 0,30 with radius: 20 and colour green
   b) centre: 0,50 with radius: 30 and colour green

6 Trim the circles 'to each other' and fillet the 'corners' with a radius of 2

7 Convert the four arcs into a single polyline with the menu bar sequence **Modify-Object-Polyline** using the **J**oin option

8 Extrude the arced polyline for a height of 6 with 0 taper, then change the colour of the extrusion to green

9 Subtract the green and blue 'holes' from the red extrusion to display the model as Fig. 32.2(e)

10 Create the following two primitives:

cylinder	*box*
centre: 45,20,−24	corner: 45,15,−24
radius: 5	length: 30
height: 6	width: 10
colour: magenta	height: 6
	colour: magenta

11 Union these two magenta primitives

12 Rectangular array the magenta cylinder/box composite:
    a) for 3 rows and 1 column
    b) row offset: 40
    c) column offset: 0

13 Mirror the three arrayed magenta composites about the points 0,0 and 0,50 and do not delete the source objects

14 Subtract the six magenta composites from the red extrusion

15 The model is now complete – Fig. 32.2(f) and the four viewport layout should be displayed as Fig. 32.3

16 Save the model layout as **MODR2004\BACKPLT**

17 Shade and use the 3D orbit command in the 3D viewport, then return the model to wire-frame representation.

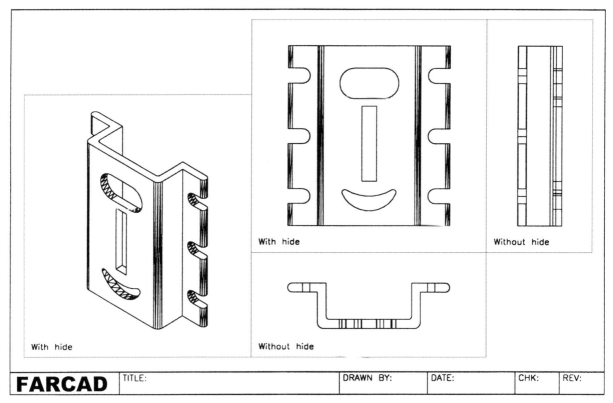

FARCAD | TITLE: | DRAWN BY: | DATE: | CHK: | REV:

Figure 32.3 Complete solid model composite of backing plate.

Investigating the model

1   Display the Inquiry toolbar and make the 3D viewport active

2   Select the LIST icon from the Inquiry toolbar and:
    prompt   Select objects
    respond   **pick the composite then right-click**
    and   AutoCAD Text Window
    with   details of the selected model
    respond   view the details then cancel the text window

3   Select the AREA icon from the Inquiry toolbar and:
    prompt   Specify first corner or [Object..
    enter   **O <R>** – the object option
    prompt   Select objects
    respond   **pick the composite**
    and   Area = 38171.70, Perimeter = 0.00

4   This is the surface area of the model

5   Select the REGION/MASS PROPERTIES icon and:
    prompt   Select objects
    respond   **pick the composite then right-click**
    prompt   AutoCAD Text Window
    with   Mass = 98283.83 and other 'technical' information about the model
    enter   **N <R>**, i.e. do not write to file

We will investigate the Mass Properties in a later chapter, but this extruded model exercise is now complete.

Chapter 33

Composite model 3 – a flange and pipe

This exercise will involve creating a composite mainly as a revolved swept primitive. Cylinder primitives will be subtracted from the revolved primitive to complete the model. As a variation, we will work with a multi-view layout instead of the model tab.

1 Open A3SOL template file as normal

2 With the lower left viewport active, restore UCS RIGHT and draw:
   a) circle: centre at 0,0 with radius: 30
   b) circle: centre at 0,0 with radius: 40
   c) line: *start point* at −200,0 and *next point* at @0,100

3 Select the REVOLVE icon from the Solids toolbar and:
   prompt   Select objects
   respond   **pick the smaller circle then right-click**
   prompt   Specify start point for axis of revolution or define axis
              [Object..
   enter   **O <R>** – the object option
   prompt   Select an object
   respond   **pick the lower end of the vertical line**
   prompt   Specify angle of revolution
   enter   **70 <R>**

4 Change the colour of the revolved pipe to green

5 Revolve the larger circle using the same entries as step 3 and change the colour of the pipe to blue

6 Erase the line then subtract the green pipe from the blue pipe

7 You may need a paper space zoom for this operation?

8 Make the lower right viewport active and restore UCS BASE

9 With the POLYLINE icon from the Draw toolbar, draw a continuous closed polyline with the following entries:
   Start point: 0,−30
   Next point: @0,−10
   Next point: @10,0
   Next point: Arc option with endpoint: @10,−10
   Next point: Line option with endpoint: @0,−50
   Next point: @30,0
   Next point: @0,70
   Next point: close option

10 Using the REVOLVE icon from the Solids toolbar:
    a) pick a point on the polyline then right-click
    b) enter **X <R>** as the axis of revolution
    c) enter **360 <R>** as the angle

11 Change the colour of the revolved solid to magenta

12 With the top left viewport active, restore USC RIGHT

13 Create a cylinder with:
    a) centre: 0,100,20
    b) radius: 25
    c) height: 30
    d) colour: number 222

14 Polar array this cylinder about the point 0,0 with the items (5) and angle to fill (360) method with rotation

15 *a*) Subtract the five cylinders from the magenta flange
    b) Union the flange and pipe

16 Centre the model with zoom-extents, then zoom to a scale of 1

17 The model is complete and is displayed in Fig. 33.1. It can now be saved as **MODR2004\FLPIPE**

18 Select the model tab, SE Isometric viewpoint, Gouraud shading then use 3D orbit. The result is impressive?

19 *Note*:
    The three composite models MACHSUPP, BACKPLT and FLPIP will be used in later chapters to demonstrate other commands, so make sure they have all be saved.

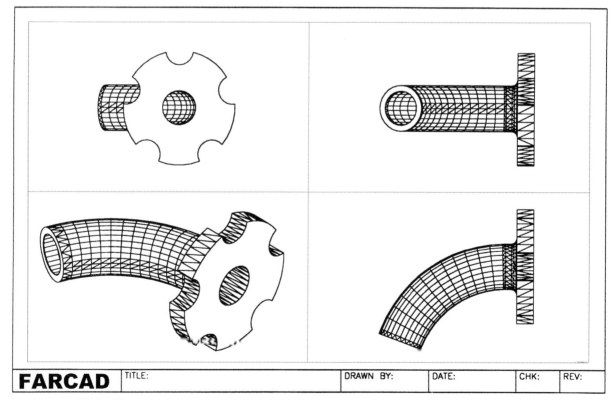

FARCAD

| TITLE: | | DRAWN BY: | DATE: | CHK: | REV: |

Figure 33.1   The completed composite model 3 – pipe/flange.

The tab layouts

Until now all models have been created using the Model tab or, occasionally, the MVLAY1 tab active. The only reason for this is that this is my preferred method of creating a solid model.

Figure 33.2 is a compilation of the four tab displays on the one sheet. This in itself is an interesting exercise, i.e. can you set the viewport configuration as Fig. 33.2 to display the FLPIP model as shown.

This completes the revolved composite exercise.

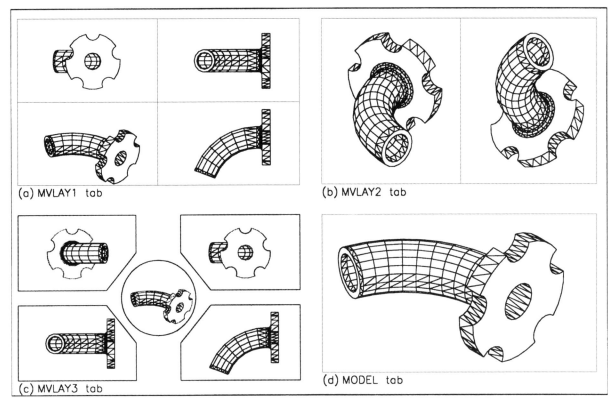

Figure 33.2   The four tab displays of the composite pipe/flange.

Chapter 34

The edge primitives

Models can be modified to include a chamfer and fillet effect. These are the edge primitives – the third type of primitive which can be created. In this chapter we will investigate how solids can be constructed with these edge primitives.

Note:
The term 'edge primitive' is not now generally used, the models being considered to have been modified or edited. I have deliberately left this chapter separate from the next chapter which investigates editing solid models.

Example 1 – a box solid

1 Open the A3SOL template file with normal settings and the MVLAY1 tab active. Refer to Fig. 34.1(a)

2 Use the BOX icon and create a primitive with:
a) corner 0,0,0
b) cube option with length: 100

3 Zoom-centre about 50,50,50 at 225 magnification

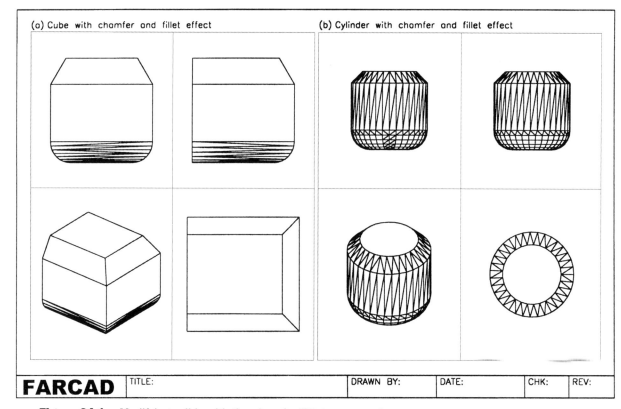

Figure 34.1 Modifying solids with the chamfer/fillet commands.

4 Select the CHAMFER icon from the Modify toolbar and:
   prompt   Select first line or [Polyline/Distance..
   respond **pick any line on top surface**
   prompt   Base surface selection
   and   *a*) one face of the cube will be highlighted
              b) it will be a 'side' or the 'top'
   prompt   Enter surface selection option [Next/OK (current)]
   respond *a*) right-click/enter if top face is highlighted
              b) enter N<R> if side face is highlighted to highlight the top face then
                 right-click/enter
   prompt   Specify base surface chamfer distance
   enter   **15 <R>**
   prompt   Specify other surface chamfer distance
   enter   **25 <R>**
   prompt   Select an edge or [Loop]
   respond *a*) pick any three sides on top surface
              b) right-click/enter

5 The top surface will be chamfered at the selected three edges

6 *Note*: entering **L** for loop will allow all edges to be chamfered with a single pick

7 Select the FILLET icon from the Modify toolbar and:
   prompt   Select first object or [Polyline/Radius/Trim]
   respond **pick any line on the base surface of the cube**
   prompt   Enter fillet radius
   enter   **20 <R>**
   prompt   Select edge or [Chain/Radius]
   respond **pick any three edges of the base then right-click**

8 The base of the cube will be filleted at the three selected edges and displayed as Fig.
   34.1(a).

Example 2 – a cylinder solid

1 Erase the cube and at the command line enter **ISOLINES <R>** and check the value is 12

2 Use the CYLINDER icon to create a cylinder with:
   a) centre: 0,0,0
   b) radius: 50
   c) height: 100

3 Zoom-centre about 0,0,50 at 200 magnification

4 Select the CHAMFER icon and:
   prompt   Select first line or..
   respond **pick the top surface circle edge**
   prompt   Base surface selection
   then   Enter surface selection option
   respond **<RETURN>** as the required surface is highlighted
   prompt   Specify base surface chamfer distance
   enter   **15 <R>**
   prompt   Specify other surface chamfer distance
   enter   **15 <R>**
   prompt   Select an edge or..
   respond **pick top circle edge then right-click/enter**

5 The top of the cylinder is chamfered with the entered values

6 Select the FILLET icon and:
   prompt   Select first object ..
   respond **pick bottom circle of cylinder**
   prompt   Enter fillet radius
   enter   **25 <R>**
   prompt   Select an edge ..
   respond **right-click** as bottom edge already selected

7 The cylinder is filleted at the base

8 The chamfer/fillet effect on the cylinder is displayed as Fig. 34.1(b).

Example 3 – The edge primitives on a solid composite

1 Erase the cylinder model and with MVLAY1 tab, UCS BASE, layer MODEL and the lower left viewport active, create four primitives with:

box	*cylinder*	*box*	*cylinder*
corner: 0,0,0	centre: 75,60,0	corner: 50,0,40	centre: 150,0,0
length: 150	radius: 40	length: 60	radius: 35
width: 120	height: 100	width: 150	height: 120
height: 100	colour: green	height: 40	colour: magenta
colour: red		colour: blue	

2 In all viewports, zoom-centre about 75,60,50 at 200 magnification in the 3D viewport and 150 magnification in the other viewports

3 Subtract the green, blue and magenta primitives from the red box and note the 'inter-penetration' effect

4 Select the FILLET icon and:
   a) pick the top circle of the green object
   b) enter a radius of 15
   c) select an edge and right-click/enter

5 The top edge of the cylinder is filleted 'outwards' and is red. Why is the fillet red and not green, as the selected object to be filleted was green?

6 Select the CHAMFER icon and:
   a) pick the top long front edge of the red box
   b) press <RETURN> if the front vertical face is highlighted or N <R> until front vertical face is highlighted then right-click/enter
   c) enter base surface distance of 10
   d) enter other surface distance of 5
   e) pick the four front edges of the blue primitive then right-click/enter
   f) failure while chamfering message displayed?

7 Repeat the CHAMFER command as step 6, but enter both chamfer distances as 5

8 The blue box primitive will then be chamfered in red – not blue?

9 Now fillet the top curve of the magenta cylinder with a radius of 20 – paper space zoom of 3D viewport recommended

10 The completed model is displayed as Fig. 34.2

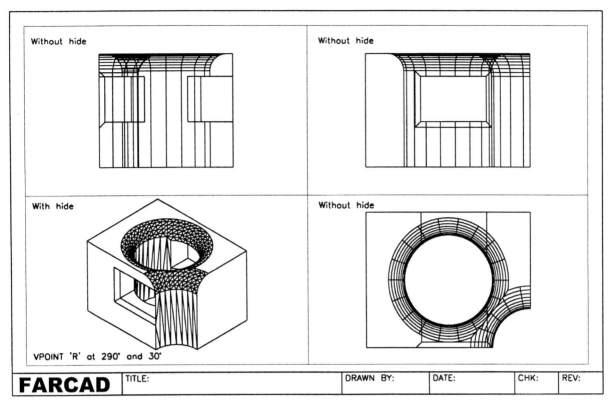

Without hide	Without hide
With hide	Without hide
VPOINT 'R' at 290° and 30°	

FARCAD	TITLE:		DRAWN BY:	DATE:	CHK:	REV:

Figure 34.2 Chamfered and filleted composite.

11 With the model tab active and a SE Isometric viewpoint, Gouraud shade and 3D orbit the model. Note the effect of the two fillet radii on the top surface

12 The model can be saved if required, but will not be used again.

A practical use for fillet/chamfer with solids

Before leaving this chapter, we will investigate a practical use of the fillet and chamfer commands with solids so:

1 Erase the composite on the screen and ensure UCS BASE, layer MODEL and lower left viewport active. Refer to Fig. 34.3

2 Create the following three box primitives:

	box1	*box2*	*box3*
corner	0,0,0	0,0,100	10,100,10
length	100	10	90
width	100	100	−10
height	10	−90	90

3 Union the three boxes and copy the unioned composite from: 0,0,0 to 100,100,0

4 Scale the copied composite about the point 100,100 by 0.75

5 Change the viewpoint in the 3D viewport with VPOINT-ROTATE and angles of 300 and 30

6 Zoom-centre in each viewport about the point 80,80,50 for 225 magnification

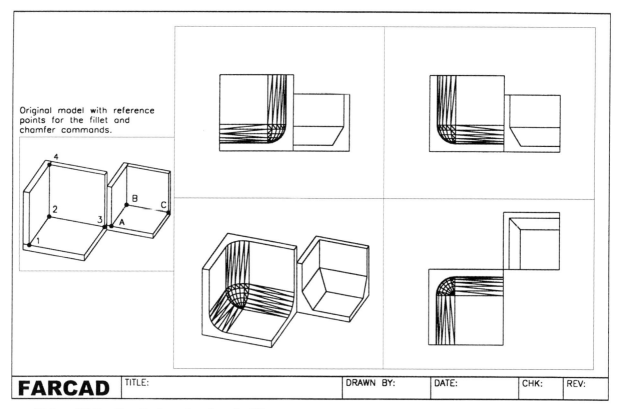

Original model with reference points for the fillet and chamfer commands.

FARCAD | TITLE: | DRAWN BY: | DATE: | CHK: | REV:

Figure 34.3 Practical use for chamfer/fillet.

7 In paper space zoom in on the 3D viewport then model space

8 Select the FILLET icon and:
   prompt   Select first object
   respond **pick edge 12**
   prompt   Enter fillet radius and enter: **25 <R>**
   prompt   Select an edge
   respond **pick edges 12,23 and 24 then right-click/enter**

9 Select the CHAMFER icon and:
   prompt   Select first line
   respond **pick edge AB**
   prompt   Enter surface selection option
   respond *a*) enter N <R> until horizontal surface is highlighted
              b) right-click/enter
   prompt   Specify base surface chamfer distance and enter: **15 <R>**
   prompt   Specify other surface chamfer distance and enter: **25 <R>**
   prompt   Select an edge
   respond **pick edges AB and BC then right-click/enter**

10 The two composites will be filleted and chamfered at the selected edges

11 *Questions*
   a) Is it possible to fillet more that three edges at the one time?
   b) Can a chamfer be added to three adjacent surfaces?

   This completes the edge primitive exercise.

Summary

1   Primitives and solids can be chamfered and filleted with the 'normal' CHAMFER and FILLET commands

2   Solids and primitives can be chamfered/filleted:
    a) inwards if a primitive
    b) outwards if a 'hole'

3   Individual edges can be chamfered and filleted

4   The chamfer command has a LOOP option allowing a complete surface to be chamfered

5   The fillet command has a CHAIN option allowing a complete surface to be filleted

6   Error messages will be displayed if the chamfer distances or the fillet radius are too large for the model being modified, and the command line will display:
    a) Failed to perform blend
    b) Failure while chamfering/Filleting.

Assignment

This activity requires a cube to be chamfered to give a 'truncated pyramid'. The model will be used in a later activity, so ensure that it is saved.

Activity 19: Penetrated pyramid of MACFARAMUS.

One of MACFARAMUS's model pyramids (well the chief digger said it was a model pyramid) was unearthed from the desert and basically consisted of four primitives:

a) a cube of side 200 with a chamfer effect
b) a cylinder with radius 25, positioned vertically through the cube centre
c) a square box of side 50, positioned horizontally through the cube centre
d) a sphere of radius 60, its centre being 30 above the centre of the top surface

You have to create this model and the suggested approach is:

1   Position the cube (red) with corner at 0,0,0

2   Position the square (blue) sided box using the Centre option

3   Position the cylinder (green)

4   Position the sphere (magenta)

5   Chamfer the cube to give a square topped pyramid, the chamfer distances being 50 for the top and 180 for the sides

6   Subtract the box, cylinder and sphere from the pyramid

7   Chamfer both open ends of the box with chamfer distances of 10

8   Fillet both ends of the cylinder with a radius of 10

9   Save the composite as **MODR2004\MODCOMP**.

Chapter 35

Solids editing

AutoCAD 2004 allows solid primitives and composites to be edited, the commands being activated from the:
a) menu bar with Modify-Solids Editing
b) Solids Editing toolbar

The editing facilities available are:

1   Boolean: union, subtraction, intersection

2   Faces: extrude, move, offset, delete, rotate, taper, color, copy

3   Edges: color, copy

4   Body: imprint, clean, separate, shell, check

The Boolean editing features have been used in the creation of the previous composites, and in this chapter we will investigate some of the other editing features.

Solids editing Example 1

1   Open your SOLA3 template file and refer to Fig. 35.1 which only displays the 3D viewport. Display the Solids, Solids Editing and Object Snap toolbars.

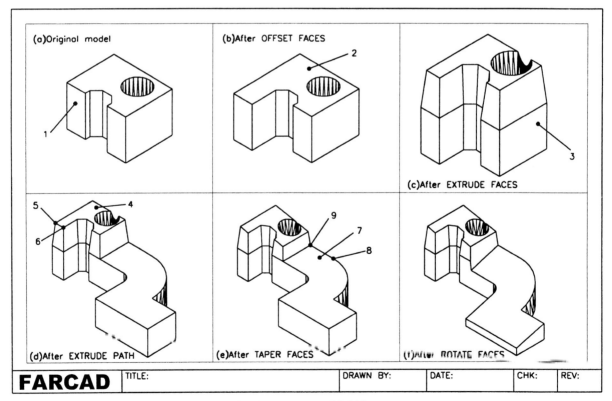

Figure 35.1   Solids editing Example 1 – 3D viewport with hide.

2 Make the Model tab active and with layer MODEL current, UCS BASE create the following:
  a) Box: corner at 0,0,0 with L: 150, W: 120, H: 100, colour: red
  b) Cylinder: centre at 100,80,0, R: 30, H: 150, colour: blue

3 Draw a polygon with eight sides, centred on 75,0,0 and inscribed in a 30 radius circle

4 Solid extrude this polygon for a height of 150 with 0 taper. Change the extruded primitive colour to green

5 Subtract the blue cylinder and green extrusion from the red box

6 Pan the composite to suit – Fig. 35.1(a)

7 Select the OFFSET FACES icon from the Solids Editing toolbar and:
  prompt Select faces or [Undo/Remove]
  respond **pick any pt1 on side surface then right-click/enter**
  prompt Specify the offset distance
  enter   **50 <R>**
  prompt Enter a face editing option and enter: **X <R>**
  prompt Enter a solids editing option and enter: **X <R>**
  and   selected face is offset as Fig. 35.1(b)

8 Select the EXTRUDE FACES icon from the Solids Editing toolbar and:
  prompt Select faces or [Undo/Remove]
  respond **pick any pt2 on the top surface then right-click/enter**
  prompt Specify height of extrusion or [Path]
  enter   **80 <R>**
  prompt Specify angle of taper for extrusion
  enter   **5 <R>**
  prompt Enter a face editing option and enter: **X <R>**
  prompt Enter a solids editing option and enter: **X <R>**
  and   top surface of model is extruded as Fig. 35.1(c)
  note   colour and taper of extruded cylinder and polygon

9 Draw a polyline using:
  Start point: 150,120
  Next point: @100,0
  arc option with arc endpoint: @150, −150
  line option with line endpoint: @100,0
  Next point: right-click/enter

10 Zoom-all and pan the model to suit

11 Menu bar with **Modify-Solids Editing-Extrude Faces** and:
  prompt Select faces or [Undo/Remove]
  respond **pick any pt3 on right face then right-click/enter**
  prompt Specify height of extrusion or [Path]
  enter   **P <R>** – the path option
  prompt Select extrusion path
  respond **pick any point on polyline**
  prompt Enter a face editing option and enter: **X <R>**
  prompt Enter a solid editing option and enter: **X <R>**

12 The selected face is extruded along the polyline path as Fig. 35.1(d)

13 Pan the model to suit

14 Select the TAPER FACES icon from the Solids Editing toolbar and:
  prompt   Select faces or [Undo/Remove]
  respond **pick any pt4 on top surface then right-click/enter**
  prompt   Specify the base point
  respond **Endpoint icon and pick pt5**
  prompt   Specify another point along the axis of tapering
  respond **Endpoint icon and pick pt6**
  prompt   Specify the taper angle and enter: **12 <R>**
  prompt   Enter a face editing option and enter: **X <R> then X <R>**

15 The selected top surface will be tapered as Fig. 35.1(e)

16 Menu bar with **Modify-Solid Editing-Rotate Faces** and:
  prompt   Select faces or [Undo/Remove]
  respond **pick any pt7 on face indicated then right-click/enter**
  prompt   Specify an axis point or [Axis by object ..
  respond **Endpoint icon and pick pt8**
  prompt   Specify the second point on the rotation axis
  respond **Endpoint icon and pick pt9**
  prompt   Specify a rotation angle
  enter   **15 <R>**
  prompt   Enter a face editing option and enter: **X <R> and X <R>**

17 The selected face is rotated about the selected side as Fig. 35.1(f)

18 Shade and 3D orbit the model then save if required

19 This exercise is now complete.

Solids editing Example 2

1 Open your SOLA3 template file and refer to Fig. 35.2. Display the Solids, Solids Editing and Object Snap toolbars.

2 With Model tab active, layer MODEL current, UCS BASE create the following two primitives:
  a) Box: corner at 0,0,0, cube option with length 200, colour red
  b) Cone: centre at 0,100,100; radius 80; Apex at @300,0 and colour green

3 Subtract the green cone from the red box and pan to suit then alter the viewpoint to suit – Fig. 35.2(a). Note that Fig. 35.2 only displays the 3D viewport.

4 The options for this example are given as a series of simple steps which you should be able to follow:
  1. *Offset faces*
    a) pick face 1
    b) distance: 75 – Fig. 35.2(b)
  2. *Rotate faces*
    a) pick face 2
    b) axis with endpoints 3 and 4 (in order given)
    c) angle 45 – Fig. 35.2(c)
  3. *Extrude faces*
    a) pick face 5
    b) height: 100 with taper: 0 – Fig. 35.2(d)
  4. *Move faces*
    a) pick face 6
    b) base point at point 7
    c) second point: @0,0,85 – Fig. 35.2(e)

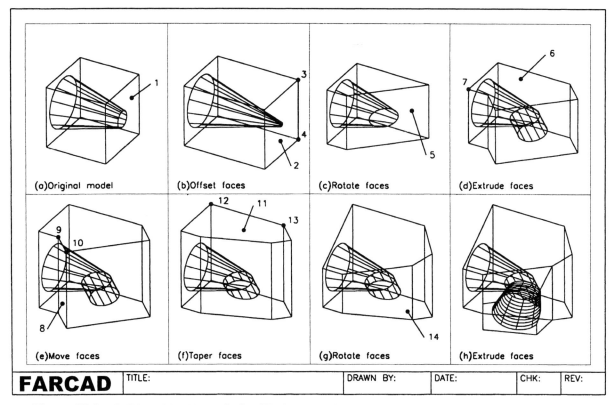

Figure 35.2 Solids editing Example 2 – 3D viewport without hide.

5. *Taper faces*
   a) pick face 8
   b) axes points 9 and 10
   c) taper angle: 65 – Fig. 35.2(f)
6. *Rotate faces*
   a) pick face 11
   b) axis with endpoints 12 and 13 (in order given)
   c) angle 35 – Fig. 35.2(g)
7. *Extrude faces*
   a) pick face 14
   b) height of extrusion: 150
   c) taper angle: 10 – Fig. 35.2(h)

This completes the second example which can be saved if required.

Note

1 The two examples have concentrated on the faces editing options

2 It is sometimes difficult to select a face for the solids editing options, and it is sometimes easier to select an edge. As an edge 'belongs' to two adjacent faces, the unwanted face can easily be removed from the 'selection' with the 'R' cntry.

3 The solids editing command allows the user repetitive options, i.e. when one option has been completed, the command is still active. The UNDO option is very useful.

4 The solids editing command is exited with two X <R> or ESC.

Solids editing Example 3

1   Open your SOLA3 template file and refer to Fig. 35.3. Display suitable toolbars.

2   With the Model tab active, layer MODEL current, UCS BASE, create the following:
    a) Box: corner at 0,0,0 with cube option with length 100
    b) Cylinder: centre at 0,0,100; radius: 50; height: 20
    c) Cylinder: centre at 100,50,100; radius: 30; height: −50
    d) Subtract the second cylinder from the box
    e) Zoom and pan to suit – Fig. 35.3(a)

3   Select the IMPRINT icon from the Solids Editing toolbar and:
    prompt   Select a 3D solid
    respond **pick the cube**
    prompt   Select an object to imprint
    respond **pick the R50 cylinder**
    prompt   Delete the source object <N>
    enter   **Y <R>**
    prompt   Select an object to imprint
    respond **<R>** – no more objects to imprint
    prompt   body editing options and enter: **X <R>**
    prompt   solids editing options and enter: **X <R>**

4   The cylinder outline is imprinted on the cube – Fig. 35.3(b)

5   With the EXTRUDE FACES option:
    a) pick face 1
    b) enter an extrusion height of 100
    c) enter a taper angle of 5 – Fig. 35.3(c)

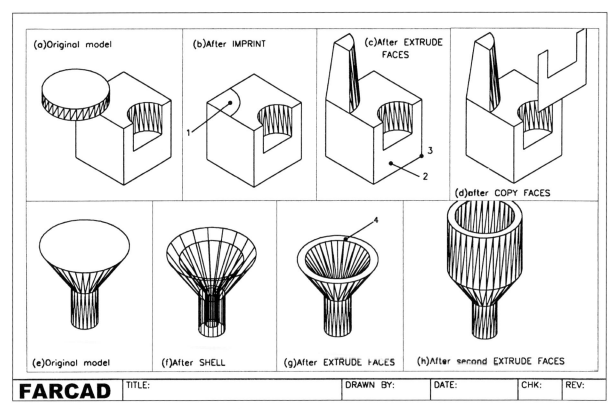

(a)Original model
(b)After IMPRINT
(c)After EXTRUDE FACES
(d)after COPY FACES
(e)Original model
(f)After SHELL
(g)After EXTRUDE FACES
(h)After second EXTRUDE FACES

FARCAD | TITLE: | DRAWN BY: | DATE: | CHK: | REV: |

Figure 35.3   Solids editing Example 3.

6 Select the COPY FACES icon from the Solids Editing toolbar and:
prompt   Select faces
respond **pick face 2 then right-click/enter**
prompt   Specify a base point or displacement
respond **Endpoint icon and pick pt3**
prompt   Specify a second point of displacement
enter   **@50,0,150 <R>** then exit the command

7 The selected face is copied as Fig. 35.3(d)

8 Erase all objects from the screen and create two primitives:
   a) cylinder: centre 0,0,0; radius: 30; height: 100
   b) cone: centre 0,0,200; base radius: 100; height: −150
   c) union the cone and cylinder – Fig. 35.3(e)

9 Select the SHELL icon from the Solids Editing toolbar and:
prompt   Select a 3D solid
respond **pick any point on the composite then right-click/enter**
prompt   Enter the shell offset distance
enter   **15<R>**
then   **X<R> and X<R>**

10 The cylinder/cone is offset by 15 in 'all directions' – Fig. 35.3(f)

11 Hide the model then regen

12 With the EXTRUDE FACES icon:
   a) pick any point on the 'top face'
   b) extrusion height: −15
   c) angle of taper: 0 – Fig. 35.3(g)

13 Repeat the extrude faces command and:
   a) pick any point 4 on the 'new rim' of the composite
   b) extrusion height: 150
   c) angle of taper: 0 – Fig. 35.3(h)

This completes the third solids editing example.

Other solids editing options

Not all of the solids editing options have been used in the worked examples. The following is a brief description of those not considered:
   a) Clean: removes all redundant edges and vertices, e.g. imprinted edges
   b) Separate: separates 3D solid objects with disjointed volumes into independent 3D objects. It **DOES NOT** separate composites created by Boolean operations into the original primitives.
   c) Check: confirms that a selected object is a valid ACIS solid
   d) Color Edges/Faces: a very useful option as it allows individual edges and faces to be coloured, i.e. a cube could be created on layer MODEL (red) and the six faces of the cube assigned different colours
   e) Copy Edges/Faces: should be obvious
   f) Delete faces: allows faces of a model to be deleted, but the option has obvious limitations. Useful with fillet/chamfer edges.

Solids editing errors

When a solids editing option is activated and completed, the command line will display:
   a) Solid validation started
   b) Solid validation completed

Solids editing may not always work due to the model selected or the option which has been activated. Typical error messages which are displayed include:

1 No solution for an edge

2 No solution for a vertex

3 No loop through new edges and vertices

4 Could not taper surface as requested

5 Improper edge/edge intersection

6 Gap cannot be filled

If an error message is obtained, then the active option cannot be performed on the selected object. Try again.

Summary

1 Solids editing allows the user several options

2 These options can result in very interesting and complex models which may be difficult to achieve from basic primitives

3 The solids editing options are divided into four categories:
   a) Boolean operations
   b) Face options
   c) Edge options
   d) Body options

4 Solid editing options which cannot be performed result in an error message being displayed

5 The solids editing command allows repetitive entries and has an undo option.

Chapter 36

Regions

A region is a *closed* 2D shape created from lines, circles, arcs, polylines, splines, etc. and can be used with the solid extrude and revolve command to create composites. When created, a region has certain characteristics:

- it is a solid of zero thickness

- it is coplaner, i.e. must be created on the one plane

- it consists of **loops** – outer and inner

- the loops must be continuous closed shapes

- every region has one outer loop

- there may be several inner loops

- inner loops must be in the same plane as the outer loop

- regions can be created with the *BOUNDARY* command

- regions can be used with the solid EXTRUDE and REVOLVE command

   Regions allow the user another method for creating solid models and very complex models can be created with regions. They can also be used when hatching is to be added to a sectioned model and when details, e.g. a true shape, is to be extracted from a model.

Example 1 – a splined shaft

1 Open your A3SOL template file as normal and display suitable toolbars

2 Ensure the DELOBJ system variable is set to 0

3 Refer to Fig. 36.1 (which only displays the 3D viewport) and create the layout from three circles having diameters 120, 40 and 16. The actual layout is your design but:
   a) use the 0,0 point as indicated
   b) polar array the 16 diameter circle

4 Zoom-centre about 0,0,40 at 200 magnification – all viewports

5 Select the SUBTRACTION icon from the Solids Editing toolbar, pick the largest circle then right-click and:
   prompt No solids or regions selected

6 Select the REGION icon from the Draw toolbar and:
   prompt Select objects
   respond **pick all circles then right-click**
   prompt 13 loops extracted
          13 Regions created

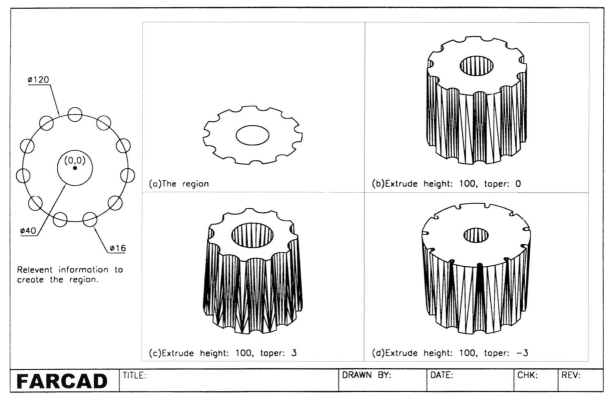

Figure 36.1   Region Example 1 – an extruded component.

7   Menu bar with **Modify-Solids Editing-Subtract** and:
   prompt   Select solids or regions to subtract from
                Select objects
   respond **pick the largest circle then right-click**
   prompt   Select solids or regions to subtract
                Select objects
   respond **pick the 12 smaller circles then right-click**
   and   the region is created as Fig. 36.1(a)

8   At this stage save as **MODR2004\REGEX** for the next exercise

9   With the lower left viewport active select the EXTRUDE icon from the Solids toolbar and:
   a) objects: pick the region
   b) height: 100
   c) taper angle: 0 – Fig. 36.1(b)
   d) hide and shade

10 Undo the hide, shade and extrude effects, then use the EXTRUDE icon with the following entries:
   a) height: 100, taper angle: 3 – Fig. 36.1(c). Undo effect
   b) height: 100, taper angle: −3 – Fig. 36.1(d)

11 This exercise does not need to be saved.

Example 2 – a revolved component

1   Open drawing file MODR2004\REGEX saved from the previous exercise with UCS BASE – Fig. 36.2(a) which again only displays the 3D viewport

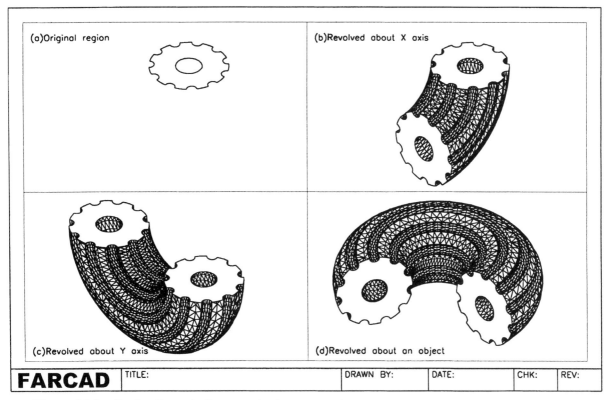

(a)Original region

(b)Revolved about X axis

(c)Revolved about Y axis

(d)Revolved about an object

FARCAD | TITLE: | DRAWN BY: | DATE: | CHK: | REV:

Figure 36.2 Region Example 2 – a revolved component.

2 Menu bar with **Tools-New UCS-Origin** and:
prompt Origin point<0,0,0>
enter **−100,−100,0 <R>**

3 Zoom-centre about 150,0,0 at 250 magnification

4 Select the REVOLVE icon from the Solids toolbar and:
prompt Select objects
respond **pick the region then right-click**
prompt Specify start point for axis of revolution or define axis by..
enter **X <R>** – the X axis option
prompt Specify angle of revolution
enter **−90 <R>**
then pan model to suit and hide – Fig. 36.2(b)

5 Undo the hide and revolve effect to leave the original region

6 Using the REVOLVE icon:
a) pick the region then right-click
b) enter Y as the axis if revolution
c) enter 180 as the angle
d) pan and hide – Fig. 36.2(c)
e) undo the hide and revolve effect

7 Draw a line from: 0,0,0 to: @0,0,100

8 Menu bar with **Modify-3D Operation-Rotate 3D** and:
   a) pick the region then right-click
   b) enter X **<R>** as the axis
   c) enter 100,100,0 as a point on the axis
   d) enter 90 as the rotation angle

9 With the REVOLVE icon:
   a) pick the rotated region the right-click
   b) enter **O <R>** – object option
   c) pick lower end of vertical line
   d) enter 240 as the angle of revolution
   e) pan to suit then hide – Fig. 36.2(d)

10 This completes the second exercise. Save if required.

Example 3 – using a boundary

1 Open your A3SOL template file as normal and refer to Fig. 36.3

2 Draw three circles:
   a) centre: 50,0, radius: 50
   b) centre: 0,50, radius: 60
   c) centre: 75,75, radius: 75

4 Make a new layer: BND, colour blue and current

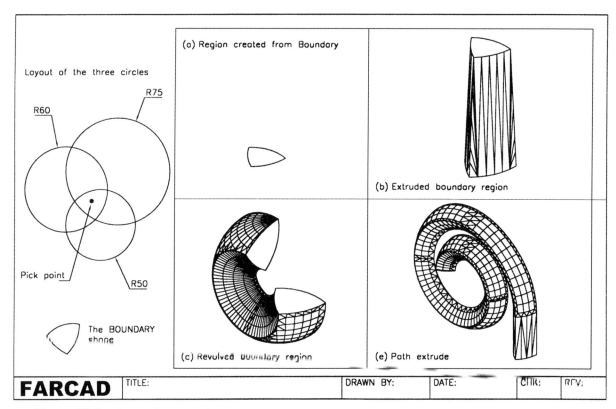

Figure 36.3   Region Example 3 – created from a boundary.

5 Zoom-centre about 50,50,50 at 200 magnification

6 Menu bar with **Draw-Boundary** and:

    prompt   `Boundary Creation dialogue box` – Fig. 36.4
    respond   1. pick Object type: Region
               2. pick Pick Points
    prompt   `Select internal point`
    respond   **pick a point indicated in Fig. 36.3**
    prompt   `Selecting everything...`
               `Selecting everything visible...`
               `Analyzing the selected data`
               `Analyzing internal islands`
    then   Select internal point
    respond   **right-click**
    prompt   `1 loop extracted`
               `1 Region created`
               `BOUNDARY created 1 region`

7 Erase the three circles to leave the blue region – Fig. 36.3(a)

8 Using the EXTRUDE icon:
    a) pick the blue boundary region then right-click
    b) enter a height of 125
    c) enter a taper angle of 2
    d) hide the model – Fig. 36.3(b)

9 Undo the hide and extrude effects to leave the blue region

10 With the REVOLVE icon:
    a) select the blue region then right-click
    b) enter Y as the axis of revolution
    c) enter 270 as the angle of revolution
    d) hide – Fig. 36.3(c)

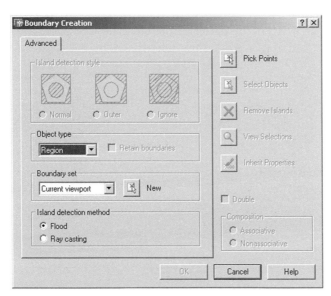

Figure 36.4   The Boundary Creation dialogue box.

11 Undo the hide and revolve effect

12 Restore UCS FRONT, and with layer MODEL current draw a polyline with:
   Start point: 0,0
   Next point: @0,100
   Next point: arc option with endpoint: @−200,0
   Next point: arc endpoint: @120,0
   Next point: arc endpoint: @−60,0 then right-click/enter

13 Restore UCS BASE and make layer BND current

14 With the EXTRUDE icon:
   a) select the blue region
   b) enter **P <R>** for the path option
   c) pick the red polyline
   d) pan to suit
   d) hide to give Fig. 36.3(d)

15 The exercise is complete, save?

16 *Note*:
   Although this exercise has been completed using the Boundary command, it could have also been completed by making the three circles into regions and then using the Boolean intersection command.

Summary

1 A region is created from closed shapes, e.g. polylines, arcs, circles, ellipses, etc.

2 Regions can be created with the BOUNDARY command

3 Regions consist of loops and all regions must have an outer loop. There can be several inner loops.

4 Regions can be extruded and revolved

5 All parts of a region are extruded/revolved to the same height or angle of revolution

6 Regions are extruded along the Z axis of the current UCS. The height of the extrusion can be positive or negative.

7 Regions can be extruded along a path.

Assignment

This activity requires a region to be created from circles, copied, scaled and then extruded to different heights.

Activity 20: Ratchet mechanism of MACFARAMUS.

While digging in a water bed outside the city of CADOPOLIS, a device was discovered which was thought to be a ratchet mechanism for a primitive type of waterwheel. This mechanism has three distinct 'parts' to it, each having the same shape.

Using the reference sizes given, create the outline from circles, convert these circles into regions, then subtract the smaller circles from the larger. The other parts of the mechanism are scaled by 0.85 and 0.55 from the original. When the three parts have been created, they have to be:

a) lower part: extruded to a height of −50 with −15 taper

b) middle part: extruded to a height of 60 with 0 taper

c) top part: revolved about a line object for −90 degrees and then polar arrayed for 3 items about the top 'circle' centre. The line for the object can be drawn at your discretion. The line I used was drawn between two endpoints of a cut out on the top surface of the second extrusion as indicated in the activity drawing.

d) *Notes*:

   1) Use a 0,0 centre point for the large circle and for copying and scaling purposes

   2) Each scaled part is positioned on top of the previous part

   3) Use your discretion for sizes not given.

Chapter 37

Inquiring into models

In this chapter we will create two new composites and then use the AutoCAD inquiry commands to determine the properties of these solids. We will also investigate how to create a material properties file.

Composite model 4 – a slip block

1   Open the A3SOL template file with normal settings, display the Inquiry toolbar and refer to Fig. 37.1

2   With Model tab active and a SE Isometric viewpoint, create the following two primitives:

Box	*Wedge*
corner: 0,0,0	corner: 120,0,0
length: 120	length: 50
width: 100	width: 100
height: 100	height: 70
colour: red	colour: blue

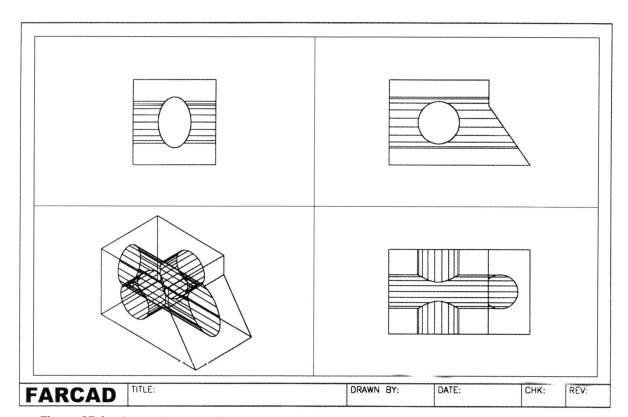

Figure 37.1   Composite model 4 – a slip block (plotted without hide).

3 Create two green cylindrical primitives with:
   a) centre: 60,0,50
      radius: 25
      centre of other end option: @0,100
   b) **elliptical** option
      centre of ellipse: 0,50,50
      axis endpoint: @20,0
      length of other axis: @0,0,30
      centre of other end: @180,0

4 *a*) union the red box and blue wedge
   b) subtract the two green cylinders from the composite
   c) note the 'curves of interpenetration'
   d) shade and note the colour effect
   e) use the 3D orbit command with the model, then restore the original 2D wire-frame representation

5 At this stage save the composite as **MODR2004\SLIPBL**

6 Select the AREA icon from the Inquiry toolbar and:
   prompt   Specify first corner point or [Object/Add/Subtract]
   enter   **O <R>** – the object option
   prompt   Select objects
   respond **pick the composite**
   prompt   Area = 98064.1, Perimeter = 0.00

7 The area value displayed is the surface area of the composite in square units (which are square mm?). A solid object has no perimeter, hence the 0 value

8 Select the REGION/MASS PROPERTIES icon from the Inquiry toolbar and:
   prompt   Select objects
   respond **pick the composite then right-click**
   prompt   AutoCAD Text Window
   with   details about the model including:
              Mass = 995878.04
              Volume = 995878.04
   prompt   Write analysis to a file?<N>
   enter   **Y <R>**
   prompt   Create Mass and Area Properties File dialogue box
   respond 1. check – Save in: named folder active
              2. check – Save as type ***.mpr** – materials properties
              3. enter File name: **SLIPBL**
              4. pick Save (more on properties file later)

9 Note that the mass and volume values are the same as AutoCAD 2004 assumes a density of 1

10 With the MVLAY1 tab active, centre the model with zoom-extents then zoom to a factor of 2, but 1.4 in the 3D viewport.

Composite model 5 – a casting block

1 Close and existing drawings then open A3SOL template file with the normal settings and:
   a) refer to Fig. 37.2
   b) zoom-centre about 37,50,18 at 150 magnification in all viewports

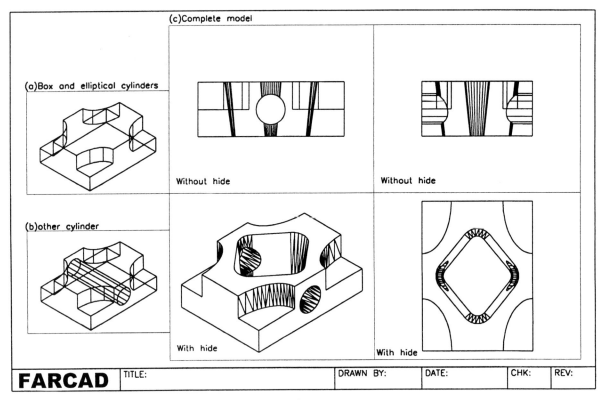

Figure 37.2 Composite model 5 – a casting block.

2 With the 3D viewport active, create a box primitive with:
 a) corner: 0,0,0
 b) length: 75, width: 100, height: 36

3 Create a **elliptical** cylinder with:
 a) centre: 0,0,36 (remember to enter C for this option)
 b) axis endpoint: @20,0
 c) length of other axis: @0,35
 d) height: −18

4 Rectangular array the elliptical cylinder:
 a) 2 rows and 2 columns
 b) row offset: 100, column offset: 75, angle of array: 0

5 Subtract the four cylinders from the box – Fig. 37.2(a)

6 Create another cylinder with:
 a) centre: 0,50,18
 b) radius: 10
 c) centre of other end option: @75,0,0

7 Subtract the cylinder from the composite – Fig. 37.2(b)

8 *a*) Draw a polyline:
  start: 0,0; next: 25,30; next: 0,60; next: −25,30; to: close
 b) Fillet the polyline with a radius of 6
 c) Move the polyline from: 0,0 by: @37.5,20
 d) Extrude the polyline for a height of 36 with −8 taper

9 Complete the model by subtracting the extruded polyline from the box composite – Fig. 37.2(c)

10 Save the model as **MODR2004\CASTBL**

11 In the MVLAY1 tab, zoom-extents then zoom to a factor of 3 in each viewport to centre the model

12 Menu bar with **Format-Units** and using the Drawing Units dialogue box alter:
   a) Length type: Engineering
   b) Precision: 0' −0.00"
   c) Drag-and-drop scale units: inches
   d) pick OK

13 Select the AREA icon from the Inquiry toolbar and:
   a) use the object option and pick the composite
   b) Area = 29416.74 square in (204.2829 square ft)

14 Select the MASS PROPERTIES icon, pick the composite and:
prompt	AutoCAD Text Window
with	Mass 155028.37 lb
	Volume 155028.37 cu in
respond	<RETURN> to the prompt
prompt	Write Analysis to a file?<N>
enter	**Y <R>**
prompt	Create Mass and Area Properties File dialogue box
respond	1. ensure file name is: CASTBL.mpr
	2. pick Save

15 *Investigate*
   a) The area and mass properties have been 'calculated' with the UCS BASE. There are two other UCS saved positions, FRONT and RIGHT. Using these saved UCS's, determine the area and mass/volume values with engineering units.
   My values were:

UCS	Area (sq in)	Mass (lb)/Volume (cu in)
BASE	29416.74	155028.37
FRONT	29416.67	155029.10
RIGHT	29416.67	155026.86

   These slight variations in mass/volume are due to the way in which AutoCAD 2004 performs the various calculations.

   b) Using the relevant commands, find the area and mass for the composite models created in previous chapters. My values with UCS BASE and decimal units were:

Model/chapter	Area	Mass
Machine support (31)	81270.75	1022333.68
Backing plate (32)	38171.70	98283.83
Pipe/flange (33)	190938.19	1305275.35

Mass Properties file

When the mass properties command is used with a solid model, the AutoCAD Text Window will display 'technical' information about the model including the mass and volume.

The user has the option of saving this information to a Mass Property file with the extension **.mpr**. As the Mass Property file is a 'text file' it can be opened in any 'text editor' type package.

To demonstrate this:

1 Select the Windows Start icon from the Windows bar at the bottom of the screen then
select **Programs-Accessories-Notepad** and:
prompt   Untitled Notepad screen
respond   1. menu bar with **File-Open**
          2. at Look in, scroll to named folder
          3. alter file name to *.mpr then <R>
          4. pick SLIPBL
          5. pick Open

2 The screen will display the saved material property file for the slip block composite –
Fig. 37.3(a)

```
        (a)Material Properties file: SLIPBL.mpr

        ----------------    SOLIDS    ----------------

     Mass:                   995878.04
     Volume:                 995878.04
     Bounding box:       X: 0.00   --   170.00
                         Y: 0.00   --   100.00
                         Z: 0.00   --   100.00
     Centroid:           X: 71.51
                         Y: 50.00
                         Z: 45.60
     Moments of inertia: X: 6600227736.14
                         Y: 10169053276.42
                         Z: 10477978687.51
     Products of inertia: XY: 3560916384.92
                          YZ: 2270735747.20
                          ZX: 2944983099.09
     Radii of gyration:  X: 81.41
                         Y: 101.05
                         Z: 102.57
     Principal moments and X-Y-Z directions about centroid:
                         I: 1943214316.07 along [0.95 0.00 -0.30]
                         J: 3004976721.16 along [0.00 1.00 0.00]
                         K: 2991525341.61 along [0.30 0.00 0.95]


        (b)Material Properties file: CASTBL.mpr

        ----------------    SOLIDS    ----------------

     Mass:                   155028.37 lb
     Volume:                 155028.37 cu in
     Bounding box:       X: 0.00   --   75.00 in
                         Y: 0.00   --   100.00 in
                         Z: 0.00   --   36.00 in
     Centroid:           X: 37.50 in
                         Y: 50.00 in
                         Z: 15.15 in
     Moments of inertia: X: 599837020.62 lb sq in
                         Y: 344914670.09 lb sq in
                         Z: 841243716.56 lb sq in
     Products of inertia: XY: 290678184.95 lb sq in
                          YZ: 117437559.14 lb sq in
                          ZX: 88078169.35 lb sq in
     Radii of gyration:  X: 62.20 in
                         Y: 47.17 in
                         Z: 73.66 in
     Principal moments (lb sq in) and X Y-Z directions about centroid:
                         I: 176681437.85 along [1.00 0.00 0.00]
                         J: 91321365.05 along [0.00 1.00 0.00]
                         K: 235664153.20 along [0.00 0.00 1.00]
```

Figure 37.3   Material Properties files from Notepad.

3   This file can then be printed if required

4   Fig. 37.3 also displays the material property file for the casting block – Fig. 37.3(b)

5   Exit Notepad to return to AutoCAD

6   *Task*: can you import the two mpr files into an AutoCAD drawing?

    The exercise is now complete.

Summary

1   Mass properties can be obtained from solid models

2   The mass properties include mass, volume, centroid, radius of gyration, etc.

3   The mass properties can be saved to an mpr text file

4   AutoCAD 2004 does not support a materials library and hence mass and volume are always the same, as density is assumed to be 1.

Assignment

MACFARAMUS designed many different types of objects including garden walls and paths. It is one of these 'garden blocks' that you now have to create as a solid model. The block was one of many used in the design of the famous walled gardens of CADOPOLIS and the block was made from the mud of the river CLYDEBER on whose banks CADOPOLIS was built.

Activity 21: Garden block of MACFARAMUS.

The creation of the garden block is straightforward:

1   Open your template file as normal

2   Using the reference sizes in the activity drawing, create the model from primitives or regions – your choice

3   Save the completed model as **MODR2004\GARDBL** as it will be used in a later chapter

4   *Task*
    a) when the model is complete, obtain the mass
    b) MACFARAMUS obtained a quantity of mud from the banks of the river CLYDEBER for making the blocks. If the mass of this mud was 1000000 units, how many blocks could he make (assuming no waste)
    c) I worked out he could create 5748 blocks.

Chapter 38

Slicing and sectioning solid models

Solid models can be sliced (cut) and sectioned relative to:
a) the XY, YZ and ZX co-ordinate planes
b) three points defined by the user
c) the viewing plane of the current viewpoint
d) user-defined objects

The two commands are very similar in operation and when used:
a) the slice command results in a new composite model. This model retains the layer and colour properties of the original solid.
b) the section command adds a 2D region to the model. The region is displayed on the current layer **BUT NO HATCHING IS ADDED TO THE REGION**.

The two commands will be demonstrated using previously created composite models.

Slice Example 1 – using the three co-ordinate planes

1 Open the composite model MODR2004\SLIPBL created in the last chapter

2 Make the Model tab active with UCS BASE and layer Model current

3 Select the SLICE icon from the Solids toolbar and:
   prompt   Select objects
   respond **pick the composite then right-click**
   prompt   Specify first point on slicing plane by [Object/Z axis/ View/XY/YZ/ZX/3 point]
   enter   **XY <R>** – the XY plane option
   prompt   Specify a point on the XY plane<0,0,0>
   enter   **0,0,50 <R>** – why these co-ordinates?
   prompt   Specify a point on desired side of plane or [keep Both sides]
   enter   **B <R>** – both sides option
   and   an XY slicing plane added to the model

4 Erase that part of the model above the slicing plane to give the Fig. 38.1(a) effect

5 Undo the erase and slice commands to restore the original model

6 Repeat the SLICE command and:
   a) pick the model then right-click
   b) enter XY as the slicing plane
   c) enter 0,0,50 as a point on the plane
   d) enter @0,0,−10 as a point on the desired side of the plane
   e) effect is the same as step 3, i.e. Fig. 38.1(a)

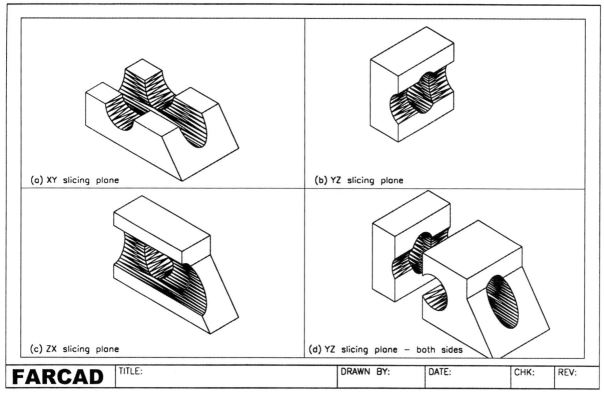

(a) XY slicing plane

(b) YZ slicing plane

(c) ZX slicing plane

(d) YZ slicing plane – both sides

FARCAD | TITLE: | DRAWN BY: | DATE: | CHK: | REV:

Figure 38.1 Slice Example 1 – using the slicing planes with model SLIPBL.

7 With the slice command, the user can either:
   a) specify a point on the desired side of the slicing plane which is to be kept
   b) keep both models on either side of the slicing plane then erase or move one of them

8 Now undo the slice effect or re-open SLIPBL

9 Menu bar with **Draw-Solids-Slice** and:
   a) pick the composite then right-click
   b) enter YZ <R> as the slicing plane
   c) enter 50,0,0 <R> as a point on the plane
   d) enter @−10,0,0 <R> to keep part 'to left' of slice plane
   e) the new composite is created as Fig. 38.1(b)

10 Undo the slice effect

11 At the command line enter **SLICE <R>** and:
   a) pick the composite then right-click
   b) enter ZX as the slicing plane
   c) enter 0,50,0 as a point on the plane
   d) enter @0,10,0 to keep part 'to right' of the slice plane
   e) new composite as Fig. 38.1(c)

12 Undo the slice effect, activate the slice command and:
   a) pick the composite then right-click
   b) select the YZ plane
   c) enter 40,0,0 as a point on the plane
   d) enter **B <R>** to keep both sides
   e) model is now 'sliced' into two new composites

13 With the MOVE command:
  a) pick the blue wedge then right-click
  b) base point: enter 170,0
  c) second point of displacement: enter @50,0
  d) the two new composites are separated – Fig. 38.1(d)

14 This completes the first exercise which does not need to be saved.

Slice Example 2 – using user-defined slicing planes

1 Open the drawing file MODR2004\MACHSUPP – the first composite solid created in Chapter 31. Make the Model tab active with USC BASE and layer MODEL current. Refcr to Fig. 38.2.

2 Activate the SLICE command and:
  prompt   Select objects
  respond **pick the composite then right-click**
  prompt   Specify first point on slicing plane ..
  respond **right-click/enter** – accepting the three point default
  prompt   Specify first point on plane and: **ENDpoint icon and pick pt1**
  prompt   Specify second point on plane and: **ENDpoint icon and pick pt2**
  prompt   Specify third point on plane and: **QUAdrant icon and pick pt3**
  prompt   Specify a point on desired side of plane ..
  enter   **@0,0,−10 <R>** – why this entry?

3 A new composite will be created as Fig. 38.2(a). This has been 'sliced' through an inclined plane.

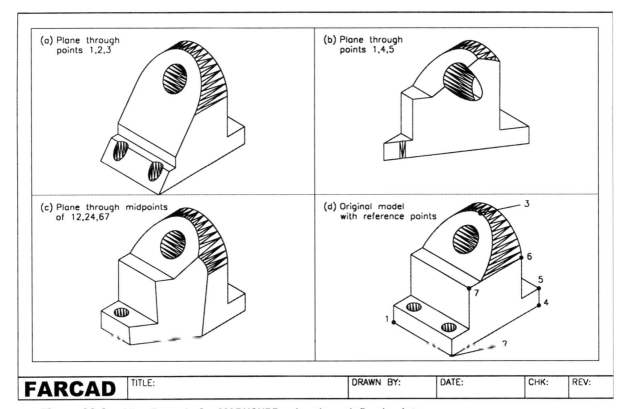

(a) Plane through points 1,2,3

(b) Plane through points 1,4,5

(c) Plane through midpoints of 12,24,67

(d) Original model with reference points

FARCAD TITLE: DRAWN BY: DATE: CHK: REV:

Figure 38.2   Slice Example 2 – MABHSUPP using three-defined points.

4 Hide and shade the model

5 Undo the shade, hide and slice commands to leave the original composite – or re-open MACHSUPP

6 With the SLICE command:
   a) pick the composite then right-click
   b) enter <R> to activate the three point option
   c) first point on plane: ENDpoint and pick pt1
   d) second point on plane: ENDpoint and pick pt4
   e) third point on plane: ENDpoint and pick pt5
   f) point on plane: enter @0,10,0
   g) hide the new composite – Fig. 38.2(b)

7 This new composite has been sliced through a vertical plane from corner to corner

8 Undo the hide and slice effects

9 Use the SLICE command on the composite, activate the three point option and:
   a) first point on plane: MIDpoint of line 12
   b) second point on plane: MIDpoint of line 24
   c) third point on plane: MIDpoint of line 67
   d) point on plane: enter: @−10,0,0
   e) new composite displayed as Fig. 38.2(c)
   f) hide and shade

10 The exercise is now complete.

Slice Example 3 – view and object options

1 Open the drawing file MODR2004\FLPIPE of the flange-pipe composite from Chapter 33

2 MVLAY1 tab active with UCS BASE and layer MODEL current

3 Refer to Fig. 38.3 and set the following 3D Views:
viewport	3D View
top left	NW Isometric
top right	NE Isometric
bottom right	SW Isometric
bottom left	SE Isometric – should be set to this

4 Re-centre about −100, −30,0 at 300 magnification and pan to suit

5 With the top left viewport active, select the SLICE icon and:
prompt	Select objects
respond	**pick the composite then right-click**
prompt	Specify first point on slicing plane or ..
enter	**V <R>** – the view option
prompt	Specify a point on the current view plane<0,0,0>
enter	**0,50,0 <R>**
prompt	Specify a point on desired side of plane ..
enter	**0, −10,0 <R>**, i.e. keep part 'behind' the plane
and	new composite displayed as Fig. 38.3(a)
then	hide the model

6 *Note*: the view option assumes a 'line of sight' which is perpendicular to the view plane

7 Undo the hide and slice effects to restore the original model

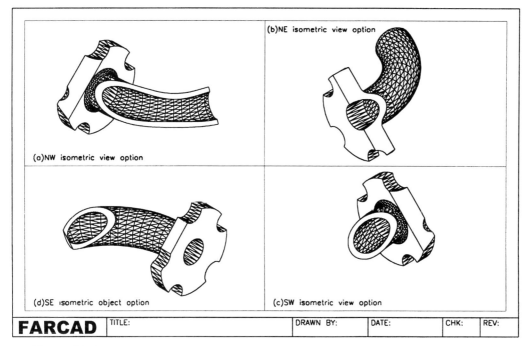

(a)NW isometric view option

(b)NE isometric view option

(d)SE isometric object option

(c)SW isometric view option

FARCAD | TITLE: | DRAWN BY: | DATE: | CHK: | REV:

Figure 38.3 Slice Example 3 – FLPIPE model with view and object options.

8 With the top right viewport active, SLICE and:
   a) pick the composite then right-click
   b) enter V <R> for the view option
   c) enter 0,50,0 <R> as a point on the view plane
   d) enter 0,−10,0 <R> as a point on the desired side on the plane
   e) new composite created
   f) hide the model – Fig. 38.3(b)

9 Undo the hide and slice effects

10 Make the lower right viewport active and slice with:
   a) the composite
   b) the view option
   c) 0,−50,0 as a point on the plane
   d) 0,10,0 as a point on the desired side of the plane
   e) new composite as Fig. 38.3(c)
   f) hide then undo both the hide and slice effects

11 Finally with the lower left viewport active:
   a) draw a circle with centre: 0,−100 and radius: 100
   b) rotate 3D this circle:
     1. about the X axis
     2. with the circle centre as a point on the axis
     3. for a reference angle of 45 degrees
   c) with the SLICE command:
     1. pick the composite then right-click
     2. enter O <R> for the object option
     3. pick the circle as the object
     4. enter 0,10,0 as a point on desired side
     5. erase the circle then hide and composite as Fig. 38.3(d)

The slice examples have now been completed.

Section example – three point option

The section command is very similar in operation to the slice command, and only one example will be demonstrated.

1 Open the drawing file MODR2004\SLIPBL from Chapter 37, MVLAY1 tab, UCS BASE, layer MODEL and the lower left viewport active. Refer to Fig. 38.4.

2 Make layer SECT current – or make a new section layer

3 Select the SECTION icon from the Solids toolbar and:
   prompt   Select objects
   respond **pick the composite then right-click**
   prompt   Specify first point on section plane by ..
   respond **<R>** – to activate the three point option
   prompt   Specify first point on plane and ENDpoint of pt1
   prompt   Specify second point on plane ENDpoint of pt2
   prompt   Specify third point on plane pick ENDpoint of pt3

4 A coloured outline as added to the model

5 Menu bar with **Tools-Inquiry-List** and:
   prompt   Select objects
   respond pick any point on the added outline then right-click
   prompt   AutoCAD Text Window
   with   details about the selected object and it is a REGION
   respond cancel the text window

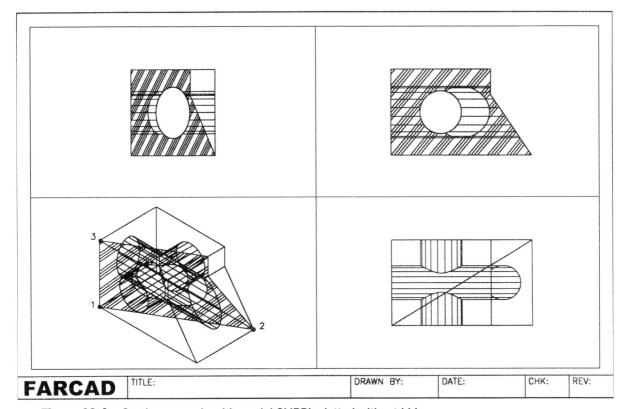

Figure 38.4   Section example with model SLIPBL plotted without hide.

6   *Note*:
    a) Hatching is not automatically added to the region
    b) Hatching must be added to the section region by the user
    c) The section region is displayed in all viewports, and the viewport specific layer concept can then be used to 'currently freeze' the section outline in specific viewports
    d) If hatching is to be added to the section region (as it should?), remember to alter the UCS position as hatching is a 2D concept

7   *Task*
    Add hatching to the added section plane using the following information:
    a) Hatching to be Predefined-ANSI-ANSI34
    b) Scale: 1 and Angle: 0
    c) Hide the model in each viewport with interesting results
    d) A new three point UCS will be required and the points should be obvious, as they are the same as the section plane points

8   In a later chapter we will section models 'more realistically'.

Using the slice and section commands

If a 'true section' has to be obtained from a solid composite, then the slice and section commands can both be used. To demonstrate this concept:

1   Open drawing file MODR2004\CASTBL from Chapter 37 with the MVLAY1 tab, UCS BASE, layer MODEL and the lower left viewport current

2   Refer to Fig. 38.5 which displays the original model in Fig. 38.5(a)

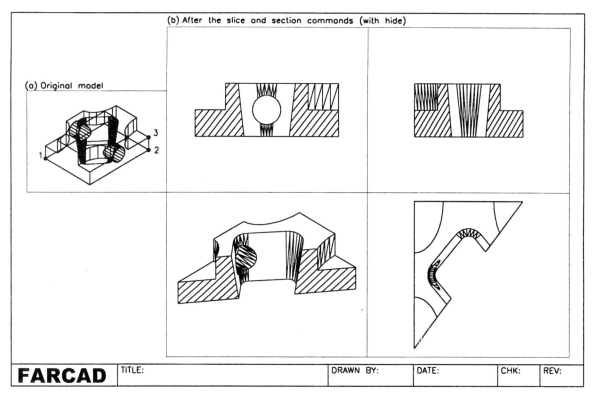

Figure 38.5   Using the SLICE and SECTION commands with CASTBL.

3   With the SLICE command:
    a) pick the composite then right-click
    b) use the three points on slicing plane option and pick the ENDpoints of points 1, 2
        and 3
    c) enter 0,100,0 as the point on the desired side on the plane
    d) a new composite will be displayed

4   With layer SECT current, use the SECTION command and:
    a) pick the new composite then right-click
    b) use the three point options with the same selection as step 3
    c) region added to composite

5   Hatch the region with correct UCS setting

6   Fig. 38.5(b) displays the result of the SLICE and SECTION commands.

Summary

1   The SLICE command produces a new composite on the same layer as the original
    model

2   The SECTION command adds a region to the section plane of the model, but does not add
    any hatching to this region. Hatching must be added by the user and is UCS dependent.

3   Both commands are very similar with the following options:
    a) the XY, YZ and ZX slicing/sectioning planes
    b) defining any three points on the slice/section plane
    c) relative to an object
    d) relative to the viewing plane in the current viewpoint

4   The SLICE command requires:
    a) a point on the slicing plane
    b) a point on the desired side of the plane which is to 'be kept'

5   The SECTION command only requires a point on the sectioning plane

6   The orientation of the UCS will affect the XY, YZ and ZX planes

7   The two commands can be activated:
    a) by icons from the Solids toolbar
    b) from the menu bar with Draw-Solids
    c) by entering SLICE or SECTION from the keyboard.

Chapter 39

Profiles and true shapes

When the HIDE command is used with a solid composite, the model is displayed with hidden line removal – as expected. From an 'engineering viewpoint' this may not be what is wanted, as the front, top and end views should have lines which represent hidden detail. AutoCAD 2004 allows this hidden detail to be 'added' to a solid model with the PROFILE command and when used, new layers are automatically added to the existing drawing.

A profile is defined as: *an image which displays the edges and silhouettes of curved surfaces of the solid for the current view*

The command will be demonstrated with previously created models.

Example 1 – slip block

1 Open drawing MODR2004\SLIPBL from Chapter 37

2 In paper space, use the LIST command and pick the top right viewport border. The AutoCAD Text Window will display information about the viewport including the **HANDLE** number. Take a note of this handle number, then repeat the LIST command selecting the other three viewport borders. In each case note the handle number as it will be referred to in the exercise. My handle numbers for the four viewports were:
   top right: 70   top left: 72
   lower right: 76   lower left: 74

3 *Note*:
   a) all objects created in AutoCAD are given a handle number and this number increases every time a new object is added to the drawing
   b) the handle numbers are in hexadecimal format – don't worry about this if you have never heard of it. It is not important at this level of AutoCAD.
   c) the handles are for assisting with the AutoCAD database
   d) your four viewport handle numbers will probably differ from mine. This is perfectly normal. Simply note them.

4 After the viewport border handle numbers have been noted, return to model space with UCS BASE

5 Make layer 0 current and the top left viewport active

6 Refer to Fig. 39.1

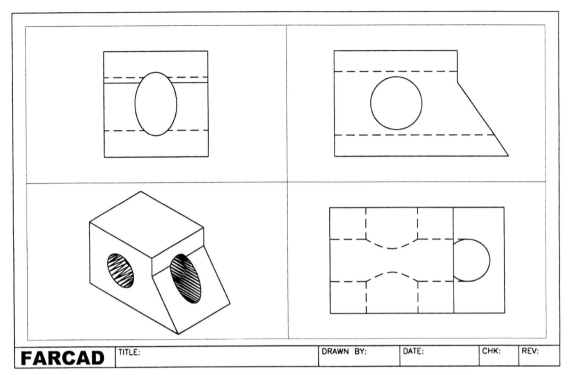

FARCAD | TITLE: | DRAWN BY: | DATE: | CHK: | REV:

Figure 39.1 Using the PROFILE command with SLIPBL.

7 Select the SETUP-PROFILE icon from the Solids toolbar and:
   prompt Select objects
   respond **pick the composite then right-click**
   prompt Display hidden profile lines on a separate layer?
   [Yes/No] <Y>
   enter **Y <R>**
   prompt Project profile lines onto a plane? [Yes/No] <Y>
   enter **Y <R>**
   prompt Delete tangential edges? [Yes/No] <Y>
   enter **Y <R>**
   and the model in the top left viewport will be displayed with black lines over the model lines

8 Menu bar with **Format-Layer** and:
   prompt Layer Properties Manager dialogue box
   with two new layers (*):
      a) **PH-72** for hidden profile lines – Hidden linetype?
      b) **PV-72** for visible profile lines – Continuous
   respond 1. check the linetype for layer PH-72
            2. if it is not Hidden, then LOAD the HIDDEN linetype and set it to layer PH-72
            3. freeze layer MODEL in the current viewport
            4. pick OK

9 The two new layers (*) will have the same handle number as the top left viewport, in my case 72

10 The top left viewport will now only display the profile visible and hidden lines for the composite as layer MODEL has been frozen in the current viewport. Do you understand why the linetype for layer PH-72 was set to HIDDEN?

11 Make the top right viewport active and from the menu bar select **Draw-Solids-Setup-Profile** and:
    prompt   Select objects
    respond **pick the composite then right-click**
    prompt   Display hidden profile lines .. and <R>
    prompt   Project profile lines .. and <R>
    prompt   Delete tangential edges.. <R>

12 The model will be displayed with black visible and hidden lines

13 Menu bar with **Format-Layer** and:
    prompt   Layer Properties Manager dialogue box
    with   Two new layers:
               a) PH-70 with HIDDEN linetype
               b) PV-70 with CONTINUOUS linetype
    respond 1. freeze layer MODEL in current viewport
               2. pick OK
    and   *a*) remember that PH-70 and PV-70 are the hidden and visible layers for my
                   viewport handles. You may have different PH and PV handle numbers.
               b) layer PH-70 has HIDDEN linetype as this linetype was loaded in step 8

14 The top right viewport will display visible and hidden lines for the model

15 Make the lower right viewport active, still with layer 0 current

16 At the command line enter **SOLPROF <R>** and:
    a) pick the model then right-click
    b) enter Y <R> to the display hidden profile lines prompt
    c) enter Y <R> to the project profile lines prompt
    d) enter Y <R> to the delete tangential edges prompt
    e) layer properties manager dialogue box and:
        1. two new layers: PH-76 and PV-76
        2. freeze layer MODEL in current viewport
        3. pick OK

17 The viewport will display the visible and hidden detail lines for the model in the viewport

18 At this stage the layout should resemble Fig. 39.1 and can now be saved if required, but not as SLIPBL

    This exercise is now complete.

Explanation

The PROFILE command is **viewport specific** and when activated:

1 Two viewport specific layers are automatically created, these being **PH-??** and **PV-??**

2 The PH layer is for hidden detail

3 The PV layer is for visible detail

4 The ?? with the layer name is the current viewport handle number and is not controlled by the user

5 The PV linetype should always be continuous

6 The PH linetype should always be hidden, but **MAY HAVE TO BE LOADED** by the user

7 The command must be used in each viewport in which profile detail has to be extracted

8 The command is generally used in viewports which display top, front and side views of the model

9 The command is generally not used in a viewport which displays a 3D view of the model

10 The hidden and visible detail added are blocks, i.e. there is a hidden detail block and a visible detail block. These blocks can be exploded if required.

Example 2 – the pipe and flange

1 Open the pipe/flange model created in Chapter 33 with MVLAY1 tab, UCS BASE, layer MODEL, lower left viewport active and refer to Fig. 39.2

2 In the top left viewport, activate the PROFILE command and:
   a) pick the model then right-click
   b) enter Y <R> to the three prompts
   c) activate the Layer Properties Manager dialogue box and:
       1. load the HIDDEN linetype and set to the new PH layer
       2. freeze layer MODEL in the current viewport

3 Repeat step 2 in the top right and lower right viewports, but do not load the HIDDEN linetype – it is already loaded?

4 Extracting profiles is as simple as this!

5 Profile drawings can be dimensioned using viewport specific layers using the same procedure as Chapter 25. I will let you try and add the dimensions for yourself, but remember to:
   a) create a new layer for each viewport
   b) set the correct UCS – dimensioning is a 2D concept

6 Save the layout when complete.

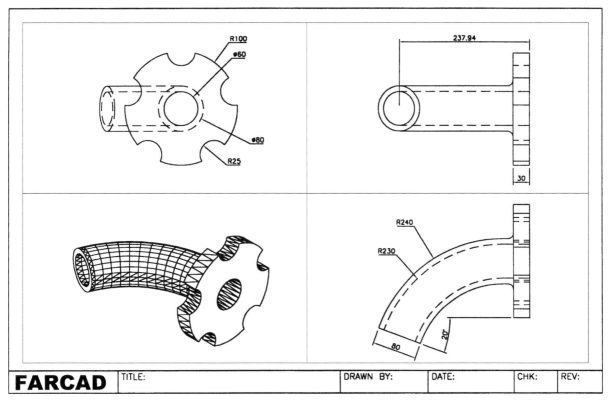

Figure 39.2 The PROFILE command with the flange/pipe model.

Profile explanation

When the profile command is used with a solid model, three prompts are displayed and it is usual to enter Y to these prompts.

a) *Display hidden profile lines on separate layers*
   This creates two blocks, one for visible lines and one for hidden lines. Two new viewport specific layers are created for this block information PV-?? and PH-??. The actual ?? number is the handle of the current viewport, i.e. it is unique. The PV (visible detail) has a continuous linetype, while the PH (hidden detail) has a hidden linetype. The hidden linetype must be loaded before it can be 'assigned' to the PH layer.

b) *Project profile lines onto a plane*
   The profile detail is displayed as 2D objects and is projected onto a plane perpendicular to the viewing direction and passing through the UCS origin

c) *Delete tangential edges*
   A tangential edge is an imaginary edge where two faces meet at a tangent. Tangential edges are not shown for most drawing applications.

True shapes

A true shape is obtained when a surface (face) is viewed at right angles. In AutoCAD 2004 this can be obtained with the PLAN and Solids Editing commands. We will use a model from a previous chapter to demonstrate how a true shape can be obtained.

1 Open the machine support model from Chapter 31 and refer to Fig. 39.3.

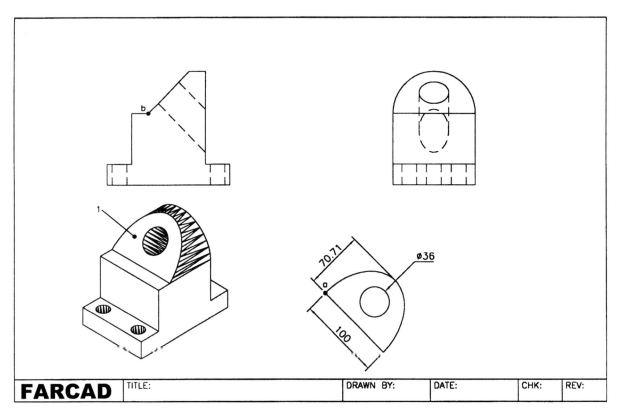

Figure 39.3 True shape extraction using composite model 1.

2 In the top two viewports, use the PROFILE command and:
   a) accept the three Y prompts
   b) freeze layer MODEL in these two active viewports
   c) load linetype HIDDEN and set to the two new PH layers

3 Create a new layer TS, colour to suit and current

4 With the lower right viewport active, restore UCS SLOPE which should have been saved with the model during the Chapter 31 exercise

5 Menu bar with **View-3D Views-Plan View-Current UCS** to display the sloped surface of the model as a 'true shape'

6 Menu bar with **Modify-Solids Editing-Copy faces** and:
   prompt   Select objects
   respond **pick any pt1 within the sloped face then right-click and enter** (toggle between the two lower viewports)
   prompt   Specify a base point and enter: **0,0 <R>**
   prompt   Specify a second point of displacement and enter: **@0,0,10 <R>**

7 *a*) freeze layer MODEL in the current (lower right) viewport to display the true shape of the sloped surface
   b) making each of the other viewports active, freeze layer TS in the other three current viewports
   c) make the lower right viewport active, still with UCS SLOPE

8 Menu bar with **Modify-Rotate** and:
   prompt   Select objects
   respond **pick any point on the true shape**
   prompt   Specify base point and enter: **0,0 <R>**
   prompt   Specify rotation angle and enter: **−45 <R>**

9 The true shape is rotated and now has to be positioned relative to the sloped surface 'from which it was copied'

10 In the lower right viewport, zoom-extents the zoom 1.5 to 'scale' the true shape to the same 'value' as the model

11 In paper space, activate the MOVE command and:
   prompt   Select objects
   respond **pick any point on the lower right viewport border then right-click**
   prompt   Specify base point or displacement
   respond **Endpoint icon and pick pt a**
   prompt   Specify second point of displacement
   respond **Endpoint icon and pick pt b**

12 Repeat the MOVE command with:
   a) objects: pick the moved viewport border
   b) base point: endpoint icon and pick pt b
   c) second point: enter @170 < −45

13 The true shape is moved to another part of the screen

14 *Task*
   a) dimension the true shape with the correct UCS and layer. This layer will need to be currently frozen in the other viewports.
   b) freeze the layer VP to give the layout as Fig. 39.3

15 The exercise is now complete and should be saved as it will be used in the next chapter.

Summary

1 Profiles can be extracted from models

2 Profiles display views of solid models with visible and hidden details

3 New layers are created with the PROFILE command, PV for visible objects and PH for hidden objects

4 The PH and PV layers are not controlled by the user and are assigned handle numbers. These handle numbers are those of the viewport in which the profile was extracted.

5 True shapes can be extracted from models with the copy faces solids editing command.

Chapter 40

Solid model dimensioning in model and paper space

All users will know that AutoCAD has two drawing environments, these being:
a) model space: used to create the model
b) paper space: used to layout the drawing sheet for plotting.

When a multiple viewport layout has been created in paper space and dimensions have to be added to the 'model', many users are unsure whether these dimensions should be added in model space or paper space. In this chapter we will investigate both model and paper space dimensioning. We discussed dimensioning with multiple viewports in Chapter 25, but will now investigate the concept further.

We will use a previously created model for the demonstration.

Getting started

1   Open the SLIPBL model from Chapter 37 with MVLAY1 tab active

2   In model space with the top left viewport active:
    a) set a FRONT 3D view
    b) zoom-extents then zoom at 1.5 scale
    c) extract a profile of the model
    d) load linetype HIDDEN if required
    e) freeze layer Model in the current viewport

3   Repeat step 2 in the top right and lower right viewports

4   You should now have the same view in three viewports

5   Refer to Fig. 40.1, restore UCS FRONT and make layer DIM current

6   Modify the 3DSTD dimension style to increase the text height to 6. This is not greatly important but makes the added dimensions 'easier to see'.

7   Linear dimension the top line of the model and diameter dimension the circle. These two dimensions will be displayed in all four viewports, as dimensioning is global.

8   Erase these two dimensions.

Viewport specific layer

1   With **Format-Layer** from the menu bar, use the Layer Properties Manager dialogue box and:
    a) pick the DIM layer then New
    b) alter the Layer1 new name to DIMTL
    c) pick Current then OK

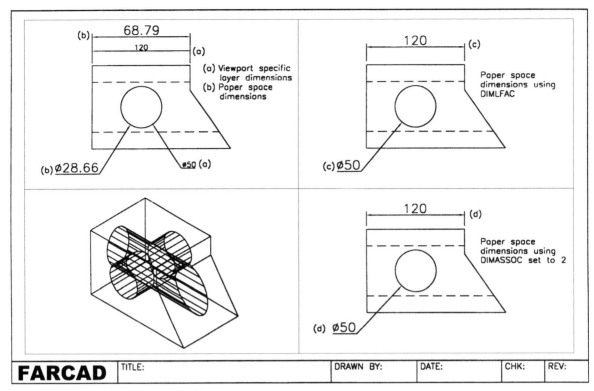

Figure 40.1   Adding dimensions in model and paper space.

2   With the top right viewport current, activate the Layer Properties Manager dialogue box and:
   a) pick the DIMTL line
   b) pick freeze in current viewport
   c) pick OK

3   Repeat step 2 in the lower left and lower right viewports, then make the top right viewport current

4   Dimension the top horizontal line and the circle as before and these two dimensions will only be displayed in the top left viewport due to the viewport specific layer DIMTL. These two dimensions are designated by (a) in Fig. 40.1.

Paper space dimensioning

1   Enter paper space

2   Dimension the same two objects as before, i.e. the top horizontal line and the circle. Use the DIM layer.

3   These two dimensions are designated by (b) in Fig. 40.1 and are obviously not correct

4   How then can paper space be used to give 'true' dimensions?

Using DIMLFAC

Dimension Linear scale Factor (DIMLFAC) is a system variable which when set to the correct value will scale linear and radial measurements and allow paper space dimensioning of model space objects to have the 'correct' value.

1   In model space make the top right viewport active, then enter paper space

2   At the command line enter **DIM <R>** and:

  prompt   Dim:
  enter   **DIMLFAC <R>**
  prompt   Enter new value for dimension variable, or Viewport
  enter   **V <R>** – the viewport option
  prompt   Select viewport to set scale
  respond   **pick the border of the top left viewport**
  prompt   DIMLFAC set to -1.74449
  then   Dim:
  respond   **ESC** to end the command line dimension sequence

3   Still in paper space, dimension the same linear and circular objects in the top right viewport. The dimensions should now be correct, designated by (c) in Fig.40.1.

Using DIMASSOC

DIMASSOC is a system variable which controls dimension associativity, i.e. it's value determines whether any added dimension will change when the object it is associated with is changed. DIMASSOC can have one of three values as follows:

  a) 0 : dimensions are displayed exploded, i.e. any part of the dimension can be selected
  b) 1 : the complete dimension is a single object and model space associativity applies. This is the normal default value.
  c) 2 : the complete dimension is a single object and paper space associativity applies

1   Paper space still active

2   At the command line enter **DIMLFAC <R>** and:

  prompt   Enter new value for DIMLFAC<-1.7445>
  enter   **1 <R>**

3   At the command line enter **DIMASSOC <R>** and:

  prompt   Enter new value for DIMASSOC<?> – probably 1 value
  enter   **2 <R>**

4   Now dimension the two objects as before using the lower right viewport. The dimensions should be correct, as (d) in Fig.40.1.

5   This exercise is now complete and need not be saved.

Paper space dimension exercise

1   Open the profile drawing saved in the previous chapter and refer to Fig. 40.2

2   With command line entry, set DIMLFAC to 1 and DIMASSOC to 2

3   Enter paper space and with layer DIM current, add the linear and circular dimensions to the two top views and the original true shape. The 70.71 true shape dimension requires some thought!

4   Are these paper space dimensions correct?

5   Enter model space with layer TS current and make viewport with the true shape active

6   Copy the true shape to another part of the viewport – you may have to 'paper space' stretch this viewport

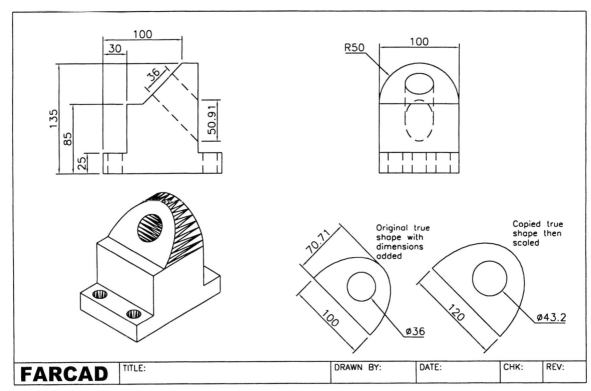

Figure 40.2   Paper space dimension exercise.

7   In paper space, with layer DIM current, add the 100 aligned dimension and the 36 diameter dimension to the copied true shape

8   Enter model space with the true shape viewport active and:
   a) activate the scale command
   b) pick the true shape and right-click
   c) pick a suitable base point
   d) enter a scale factor of 1.2

9   The model space shape **and the paper space dimensions** should both be scaled by 1.2

10 This is **true associativity**, i.e. paper space dimensions are associated with model space objects

11 In model space, move the copied true shape and the dimensions should move as well?

   This exercise is complete and can be saved.

Summary

1   Dimension can be added to models in both model and paper space

2   Multiple viewport model space dimensioning requires viewport specific layers to be made by the user

3   Paper space dimensions can be added to multiple viewport models by setting the DIMLFAC system variable

4   True associative paper space dimensions can be added to models in a multiple view-port layout with the DIMASSOC system variable set to 2. **This is the recommended method**.

Assignment

This activity requires the garden block from the previous activity to be dimensioned.

Activity 22: Adding dimensions to the garden block of MACFARAMUS.

1   Open your Activity 21

2   Set DIMASSOC to 2

3   Extract profiles from the top and front views then currently freeze the model layer in these viewports

4   Add suitable paper space dimensions to the model layout

5   Extract a diagonal section through the block and hatch using the AR-CONC predefined pattern at a suitable scale, then add the dimensions as shown (MACFARAMUS was not aware that it was considered bad practice to dimension a section view)

6   Save with a suitable name.

Chapter 41

A detailed drawing

In this example a new composite will be created and used to display the model as a detailed drawing. The model to be created is a desk tidy, so open your A3SOL template file with MVLAY1 tab active, UCS BASE, layer MODEL, lower left viewport active.

Altering the viewports

1 In paper space select the STRETCH icon from the Modify toolbar and:
   prompt Select objects
   enter **C <R>** – the crossing option
   prompt Specify first corner and enter: **160,100 <R>**
   prompt Specify opposite corner and enter: **260,200 <R>**
   prompt 4 found, Select objects
   respond **right-click**
   prompt Specify base point or displacement and enter: **200,135 <R>**
   prompt Specify second point of displacement and enter: **@−25,25 <R>**

2 The viewport configuration will be altered

3 Return to model space with UCS BASE.

The basic shape

1 Refer to Fig. 41.1

2 With the lower left viewport active restore UCS FRONT

3 Draw a polyline using the following entries:
   Start point: 0,0   *Next point*: @156,0   *Next point*: @0,15
   Next point: @−132,0   *Next point*: @0,10   *Next point*: @−24,0
   Next point: close

4 With the EXTRUDE icon from the Solids toolbar, extrude the red polyline with:
   a) height: **−85**
   b) taper: **0**

5 Restore UCS BASE and zoom-centre about the point 78,42,8 at **0.9XP** in the 3D viewport and **1XP** in the other three viewports. The XP entry is to allow for the different viewport sizes, i.e. the model is being zoomed about a centre point relative to the size of the viewport.

6 The extruded polyline will be displayed as Fig. 41.1(a).

The top

1 Still with UCS BASE, create a box primitive with:
   a) corner: 0,12,25
   b) length: 6; width: 61; height: 5

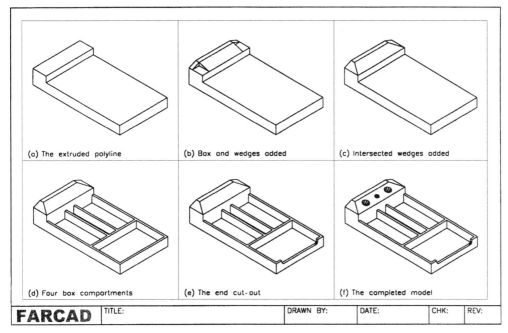

(a) The extruded polyline

(b) Box and wedges added

(c) Intersected wedges added

(d) Four box compartments

(e) The end cut-out

(f) The completed model

FARCAD | TITLE: | DRAWN BY: | DATE: | CHK: | REV:

Figure 41.1 Steps in the construction of the desk tidy (3D view only with hide).

2 Create three wedges using the following information:

	corner	length	width	height
wedge 1	6,12,25	18	61	5
wedge 2	6,12,25	12	−6	5
wedge 3	6,73,25	12	6	5

3 With the 2D ROTATE icon, rotate the following wedges:
   a) wedge 2 about the point 6,12,25 by **−90**
   b) wedge 3 about the point 6,73,25 by **90**

4 Union the box and the three wedges with the red extrusion

5 The model at this stage resembles Fig. 41.1(b)

6 Create another two wedges with:

	corner	length	width	height
wedge 4	6,0,25	18	12	5
wedge 5	24,12,25	12	−18	5

7 Rotate wedge 5 about the point 24,12,25 by **−90**

8 Menu bar with **Modify-Solids Editing-Intersect** and:
   prompt Select objects
   respond **pick wedges 4 and 5 then right-click**

9 Menu bar with **Modify-3D Operation-Mirror 3D** and:
   prompt Select objects
   respond **pick the intersected 4–5 wedges then right-click**
   prompt Specify first point of mirror plane or ..
   enter   **ZX <R>** – the ZX plane option
   prompt Specify point on ZX plane
   enter   **0,42.5,25 <R>**
   prompt Delete source objects?<N> and enter: **N <R>**

10 Now union the two sets of intersected wedges with the composite and the model should be displayed as Fig. 41.1(c).

The compartments

1 The desk tidy compartments will be created from boxes subtracted from the composite

2 With lower left viewport active and UCS BASE, create the following four box primitives:

	box1	box2	box3	box4
corner	153,3,3	100,3,3	100,36,3	100,52,3
length	−50	−76	−76	−76
width	79	30	13	30
height	20	20	20	20
colour	magenta	blue	green	blue

3 Subtract the four boxes from the red composite – Fig. 41.1(d)

4 Shade the model in the 3D viewport then return to 2D wire-frame representation.

The end cut-out

1 Lower left viewport active with UCS BASE

2 Set a new UCS position with the three point option using:
   a) origin: 156,0,0
   b) X axis position: 156,85,0
   c) Y axis position: 156,0,15
   d) save UCS position as NEWEND

3 Draw a polyline with the following keyboard entries:
   Start point: 10,15
   Next point: @0,−3
   Next point: arc option with arc endpoint: @3,−3
   Next point: line option with line endpoint: @59,0
   Next point: arc option with arc endpoint: @3,3
   Next point: line option with line endpoint: @0,3
   Next point: close

4 Set ISOLINES to 6 and FACETRES to 0.5

5 Extrude the polyline for a height of **−3** with **0** taper

6 Subtract the extruded polyline from the composite – Fig. 41.1(e)

7 Restore UCS BASE.

The holes on the slope

1 In paper space zoom-in on the 3D sloped area then model space

2 Set a new UCS position with the three point option using:
   a) origin: 24,42.5,25
   b) X axis position: 24,85,25
   c) Y axis position: 6,42.5,30
   d) save UCS position as SLOPE

3 Create three cylinders, colour magenta with:

	cylinder 1	cylinder 2	cylinder 3
centre	0,9.34077,0	20,9.34077,0	−20,9.34077,0
radius	3	5	5
height	−12	−20	−20

4　Subtract the three cylinders from the red composite – Fig. 41.1(f)

5　In paper space, zoom-previous and return to model space.

The complete model

1　Restore UCS BASE

2　The model is now complete and your screen display should be similar to Fig. 41.2

3　Save the model at this stage as **MODR2004\DESKTIDY**

4　With the model tab active use the 3D orbit command with Gouraud shading then return the original view as wire-frame.

Extracting a profile

1　With the MVLAY1 tab and the top left viewport active, select the SETUP PROFILE icon from the Solids toolbar and:
　prompt　Select objects
　respond　**pick the composite then right-click**
　prompt　Display hidden profile lines .. and enter: **Y <R>**
　prompt　Project profile line .. and enter: **Y <R>**
　prompt　Delete tangential edges and enter: **Y <R>**

2　The model display black lines in the active viewport

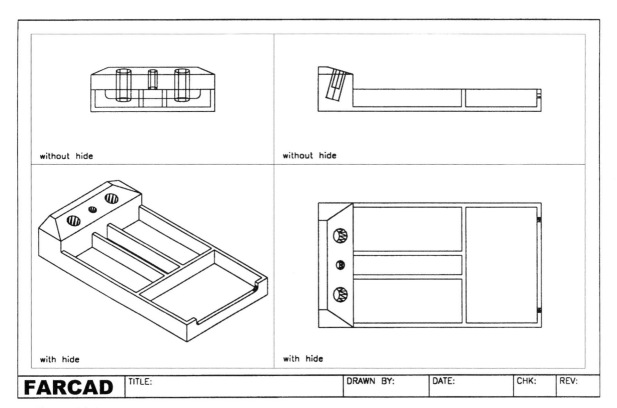

Figure 41.2　Complete desk tidy model.

3   Using the Layer Properties Manager dialogue box:
    a) note new layers PH?? and PV??
    b) load the linetype HIDDEN (if required) and set the new PH?? layer with this linetype
    c) freeze layer MODEL in this active viewport.

Extracting the section

1   With the top right viewport active, ensure UCS BASE and make layer SECT current

2   With the SECTION icon from the Solids toolbar:
    a) pick the composite then right-click
    b) enter ZX as the section plane
    c) enter 0,42,5,0 as a point on the plane
    d) a region will be displayed in all viewports

3   *a*) currently freeze layer MODEL in the top right viewport
    b) currently freeze layer SECT in the other three viewports

4   With an appropriate UCS setting, add hatching to the region using: pattern name: ANSI32; scale: 1; angle: 0.

Extracting the true shape

1   Refer to Fig. 41.3

2   The layout on your screen at present will differ from Fig. 41.3 so in paper space use the MOVE command to interchange the viewports, i.e. the two lower viewports to the top of the paper sheet and the two top viewports to the bottom of the sheet. Use the endpoint icon and pick a viewport border corner.

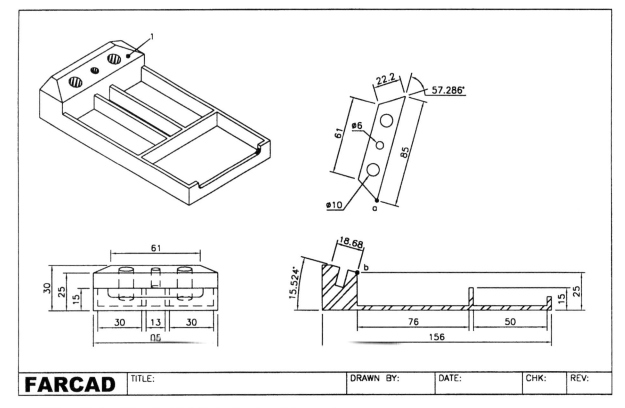

Figure 41.3   Extracting details for the desk tidy model.

3   Make a new layer named TS, colour to suit and current

4   In the top left and the two lower viewports, currently freeze the new TS layer

5   With the top right viewport active, restore UCS SLOPE

6   Menu bar with **Modify-Solids Editing-Copy faces** and:
    prompt   Select faces
    respond **pick any point within the sloped face indicated by pt1 in the top left viewport then right-click/enter**
    prompt   Specify a base point and enter: **0,0,0 <R>**
    prompt   Specify a second point and enter: **@0,0,0 <R>**
    prompt   Enter a face editing option and enter: **X <R>** then **X <R>**

7   The face has been copied 'on to itself'

8   Now currently freeze layer MODEL in the top right viewport to display the coloured true shape of the sloped surface

9   Menu bar with **View-3D Views-Plan View-Current UCS** to display the copied face (true shape) in plan view. The position of this plan view may not be ideal.

10 With the 2D ROTATE icon from the MODIFY toolbar:
    a) pick the shape and **<R>**
    b) enter 0,0 as the base point
    c) enter **74.476** as the rotation angle – this angle should become apparent once the dimensions have been added

11 Pan the rotated shape to a suitable part of the viewport

12 In paper space activate the MOVE command and:
    a) objects: pick the top right viewport border then right-click
    b) base point: endpoint icon and pick pt a
    c) second point: endpoint icon and pick pt b

13 Repeat the MOVE command and:
    a) objects: pick the same viewport border then right-click
    b) base point: endpoint icon and pick pt b
    c) second point: enter **@50<74.476 <R>**

14 The true shape is now positioned relative to the slope from which it was copied.

Task

1   Inquire into the model and:
    a) Area: 47433.58
    b) Mass: 106339.71

2   Modify the dimension style and set the decimal angle precision to 0.000

3   With DIMASSOC set to 2, add paper space dimensions to the layout as Fig. 41.3

4   The detail exercise is now complete and can be saved.

5   This detailed drawing has used most of the solid model concepts we have investigated, i.e. profile extraction, section extraction, paper space dimensioning, etc.

Chapter 42

Blocks, wblocks and external references

Solid model blocks and wblocks can be created and inserted into drawings like any other 2D or 3D object. In this chapter we will create two interesting (I hope) solid model assembly drawings from blocks and wblocks and then investigate solid model external references. We will also investigate the 'interference' between solids.

Block example – a desk tray assembly

1   Open your A3SOL template file and make the MVLAY1 tab active

2   In all viewports, zoom-centre about 55,40,20 at 0.75XP

3   Set the ISOLINES system variable to 8 and FACETRES to 0.5

4   Refer to Fig. 42.1.

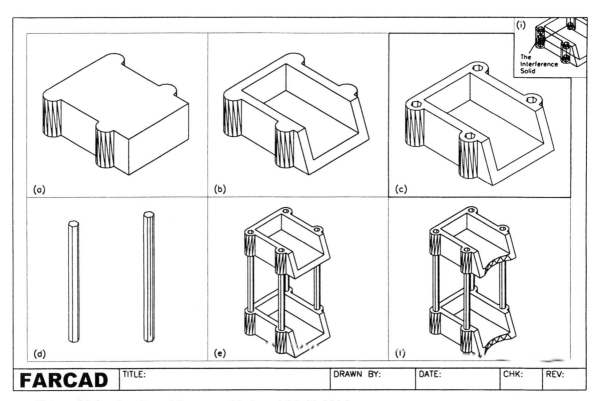

Figure 42.1   Creation of the assembled model (with hide).

The Tray

1 Make a new layer TRAY, colour blue and current

2 With the 3D viewport current and UCS BASE, create two primitives from:
    Box   *Cylinder*
    corner: 0,0,0   centre: 10,0,0
    length: 110   radius: 10
    width: 80   height: 40
    height: 40

3 Rectangular array the cylinder:
    a) for 2 rows and 2 columns
    b) row offset: 80 and column offset: 70
    c) angle of array: 0

4 Union the box and the four cylinders – Fig. 42.1(a)

5 Create another two primitives from:
    Box   *Wedge*
    corner: 10,10,10   corner: 110,0,40
    length: 110   length: −15
    width: 60   width: 80
    height: 40   height: −40

6 Subtract the box and wedge from the composite – Fig.42.1(b)

7 Draw a polygon with:
    a) sides: 6
    b) centre: 10,0,0
    c) inscribed in a circle of radius 5

8 Solid extrude the polygon for a height of 10 with 0 taper

9 Menu bar with **Modify-3D Operation-3D Array** and:
    prompt   Select objects
    enter   **L <R><R>** – two returns for last object (extruded polygon)
    prompt   Enter the type of array and enter: **R <R>**
    prompt   Enter the number of rows and enter: **2 <R>**
    prompt   Enter the number of columns and enter: **2 <R>**
    prompt   Enter the number of levels and enter: **2 <R>**
    prompt   Specify the distance between rows and enter: **80 <R>**
    prompt   Specify the distance between columns and enter: **70 <R>**
    prompt   Specify the distance between levels and enter: **30 <R>**

10 Subtract the eight cylinders from the composite – Fig. 42.1(c). A paper space zoom may be needed to help with this, but remember to return to model space.

11 At the command line enter **-BLOCK <R>** and:
    prompt   Enter block name and enter: **TRAY <R>**
    prompt   Specify insertion base point and enter: **0,0,0 <R>**
    prompt   Select objects
    respond   **pick the composite then right-click**

12 The tray block may disappear. If it does not, erase it. Remember the DELOBJ system variable?

13 *Note*:
The -BLOCK entry allows the prompts to be entered from the keyboard and not from the dialogue box.

The support

1 Make a new layer SUPPORT, colour magenta and current

2 Draw two polygons then solid extrude using the following:

	polygon 1	polygon 2
a) sides:	6	6
b) centre:	0,0,0	50,50,0
c) inscribed:	radius 5 circle	radius 5 circle
d) height:	130	140
e) taper:	0	0

3 Change the colour of the 140 high extrusion to green

4 The two extrusions are displayed as Fig. 42.1(d)

5 Make two blocks of these supports with:
a) Name: SUP1
   Insertion base point: 0,0,0
   Objects: pick the red extrusion and <RETURN>
b) Name: SUP2
   Insertion base point: 50,50,0
   Objects: pick the green extrusion and <RETURN>.

Inserting the blocks

1 Zoom-extents the zoom 1.2 in all viewports

2 Make layer TRAY current, UCS BASE and the lower left viewport active

3 Menu bar with **Insert-Block** and:
prompt Insert dialogue box
respond 1. at Name scroll and pick TRAY
        2. ensure the on-screen prompts are active (tick)
        3. ensure Explode not active (no tick)
        4. pick OK
prompt Specify insertion point
enter **0,0,0 <R>**
prompt Enter X scale factor and enter: **1 <R>**
prompt Enter Y scale factor and enter: **1 <R>**
prompt Specify rotation angle and enter: **0 <R>**

4 Repeat the INSERT command and insert block TRAY:
a) at the point 0,0,150
b) full size (i.e. X=Y=1) with 0 rotation

5 At the command line enter **-INSERT <R>** and:
prompt Enter block name and enter: **SUP2 <R>**
prompt Specify insertion point and enter: **10,0,30 <R>**
prompt Enter X scale factor and enter: **1 <R>**
prompt Enter Y scale factor and enter: **1 <R>**
prompt Specify rotation angle and enter: **0 <R>**

6 Menu bar with **Modify Solids Editing-Union** and:
prompt Select objects
respond **pick the three inserted blocks then right-click**
prompt At least 2 solids or coplaner regions must be selected

7 What does this prompt mean?

8 Menu bar with **Tools-Inquiry-List** and select any one of the inserted objects. The AutoCAD text window will be displayed with information about the selected object and it will be 'defined' as a *Block Reference*. Thus inserted blocks cannot be used with the three Boolean operations if they have not been exploded.

9 Cancel the text window and using the EXPLODE icon from the Modify toolbar, select the three inserted blocks

10 Use the LIST command again and select any one of the exploded blocks. It will be 'defined' as a 3D solid. Cancel the text window.

Checking for interference

1 Make layer 0 current

2 Menu bar with **Draw-Solids-Interference** and:
  prompt Select the first set of solids
           Select objects
  respond **pick the top tray then right-click**
  prompt Select the second set of solids
           Select objects
  respond **pick the support then right-click**
  prompt Comparing 1 solid against 1 solid
           Interfering solids (first set): 1
                           (second set): 1
                   Interfering pairs: 1
           Create interference solids>[Yes/No] <N>
  enter   **Y <R>**

3 The model will be displayed with a black solid (layer 0) where interference occurs between the tray and the support leg. This interference is due to the support leg (SUP2) being too long, or the TRAY having been inserted into the drawing at the wrong insertion point. We deliberately created the support leg (SUP2) with a height of 140 to obtain interference.

4 The interference effect is displayed in Fig. 42.1(i)

5 Erase the support leg and the black interference hexagonal prism will still be displayed as it is itself an object. Erase this interference prism.

Inserting the correct support

1 Still with the two (exploded) inserted trays displayed?

2 UCS BASE, lower left viewport active and layer SUPPORT current

3 INSERT the support block SUP1 with:
  a) insertion point: 10,0,30
  b) full size with 0 rotation
  c) explode this inserted block

4 Rectangular array the inserted support:
  a) for 2 rows and 2 columns
  b) row offset: 80 and column offset: 70
  c) angle of array: 0

5 Make layer MODEL current

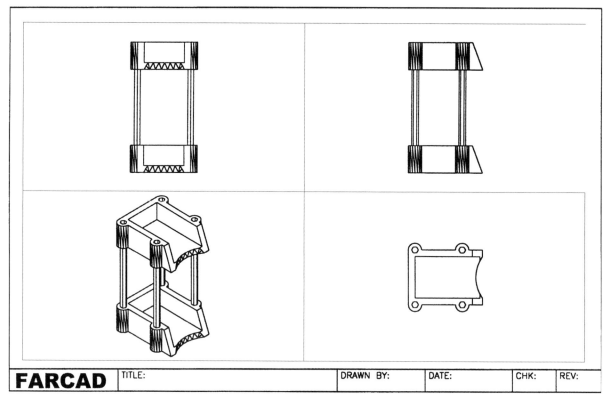

FARCAD | TITLE: | | DRAWN BY: | DATE: | CHK: | REV:

Figure 42.2   Desk tray assembly model created from blocks.

6   Menu bar with **Modify-Solids Editing-Union** and pick the six exploded blocks then right-click – Fig. 42.1(e)

7   The model at this stage is displayed in Fig. 42.1(e) without hide.

Completing the model

1   Create a cylinder with:
    a) centre: 150,40,0
    b) radius: 50 and height: 200

2   Subtract this cylinder from the composite – Fig. 42.1(f)

3   The model is now complete and can be saved as **MODR2004\DESKTRAY**

4   Figure 42.2 displays the four viewport configuration of the completed model assembly.

Block assembly example – a wall clock

In this example seven different coloured objects will be created and saved as blocks for an assembly drawing. The assembly will then be used for profile and section extraction.

1   Open your A3SOL template file with the MODEL tab active

2   Refer to Fig. 42.3 which displays the objects to be created

3   Restore UCS FRONT with layer MODEL current.

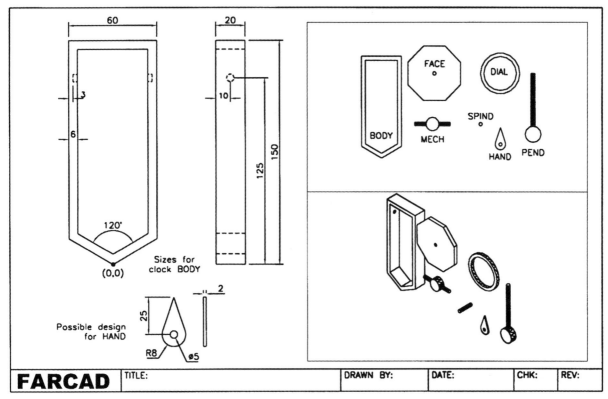

Figure 42.3   Wall clock solids for block creation.

Creating the blocks

Body (red)

1   Create the clock body from two polylines using the sizes given. The start point should be at 0,0 and your discretion is needed for any sizes not given. (I created the body from lines then used the polyline edit command to join them into one polyline. I then offset the polyline by 6).

2   Solid extrude the two polylines for a height of −20 with 0 taper

3   Subtract the inside extrusion from the outside extrusion

4   Set ISOLINES to 6 and FACETRES to 0.5

5   Create a cylinder with:
    a) centre: −27,125,−0 and diameter: 5
    b) centre of other end option: @54,0,0

6   Subtract the cylinder from the extruded composite

7   Make a block of the composite body with:
    a) block name: BODY
    b) insertion base point: 0,0,0.

Face (blue)

1   Draw a blue octagon with:
    a) centre: 80,125,0
    b) circumscribed in a circle of radius: 40

2   Solid extrude the octagon for a height of 6 with 0 taper

3   Create a cylinder with:
   a) centre: 80,125,0
   b) diameter: 5
   c) centre of other end option at: @0,−15
   d) colour: blue

4   Subtract the cylinder from the extruded blue octagon

5   Make a block of the extruded composite with:
   a) block name: FACE
   b) insertion base point: 80,125,0.

Dial (green)

1   Create two green cylinders:
   a) centre: 180,125,0; diameter: 60; height: 5
   b) centre: 180,125,0; diameter: 50; height: 5

2   Subtract the smaller cylinder from the larger cylinder

3   Block the composite as DIAL with insertion base point: 180,125,0.

Mechanism (magenta)

1   Create two magenta cylinders:
   a) centre: 50,50,0; diameter: 5; centre of other end: @54,0,0
   b) centre: 77,50,−5; diameter: 20; height: 10

2   Union the two cylinders

3   Block the composite as MECH, insertion base point: 77,50,0.

Pendulum (cyan)

1   Create two cyan cylinders:
   a) centre: 230,125,0; diameter: 6; centre other end: @0,−90,0
   b) centre: 230,35,−5; diameter: 24; height: 10

2   Union the two cylinders

3   Block the composite as PEND with insertion base point: 230,125,0.

Spindle (colour number 124)

1   Create a cylinder (colour number 20) with:
   a) centre: 150,50,0
   b) diameter: 5 and height: 25

2   Block as SPIND with insertion base point: 150,50,0.

Hand (colour number 124)

1   Create a cylinder with centre: 180,20,0, diameter: 5 and height: 2

2   Create a polyline outline using sizes as a guide – own design

3   Solid extrude the polyline for a height of 2 with 0 taper

4   Subtract the cylinder from the extrusion

5   Block as HAND with insertion base point: 120,20,0.

Inserting the blocks

1   Still with UCS FRONT, layer MODEL and MODEL tab active

2   Erase any objects still displayed

3   At the command line enter **-INSERT <R>** and:
    prompt   Enter block name and enter: **BODY**
    prompt   Specify insertion point and enter: **0,0,0**
    prompt   Enter X scale factor and enter: **1**
    prompt   Enter Y scale factor and enter: **1**
    prompt   Specify rotation angle and enter: **0**

4   Pan the inserted block to the lower centre of the screen

5   Insert the other blocks using the following information:

name	insertion point	scale	rot
FACE	0,125,0	X=Y=1	0
DIAL	0,125,6	X=Y=1	0
MECH	0,125,−10	X=Y=1	0
PEND	0,125,−10	X=Y=1	0
SPIND	0,125,−5	X=Y=1	0
HAND	0,125,15	X=Y=1	0
HAND	0,125,18	X=Y=0.75	−150

6   *a*) Gouraud shade the model with the model tab active
    b) use the 3D orbit command – impressive result?
    c) return the model to 2D wire-frame representation

7   With the MVLAY1, zoom-extents the zoom to a factor of 1.6 in all viewports

8   **DO NOT** union the inserted blocks but save as **MODR2004\CLOCK** for the rendering exercises.

Tasks

1   Union the eight inserted blocks – explode needed?

2   In the lower right viewport extract a profile of the model to display hidden detail and remember:
    a) hidden linetype must be loaded and set to the pH layer
    b) currently freeze layer MODEL in the viewport
    c) optimise the LTSCALE variable

3   In the top left viewport extract a vertical section through the model and remember:
    a) use the layer SECT with the required section plane
    b) add hatching to the region – UCS is important
    c) currently freeze layer MODEL in this viewport
    d) currently freeze layer SECT in other viewports

4   Add an additional viewport to display the assembled model from below

5   The final layout should be as Fig. 42.4 and can be saved.

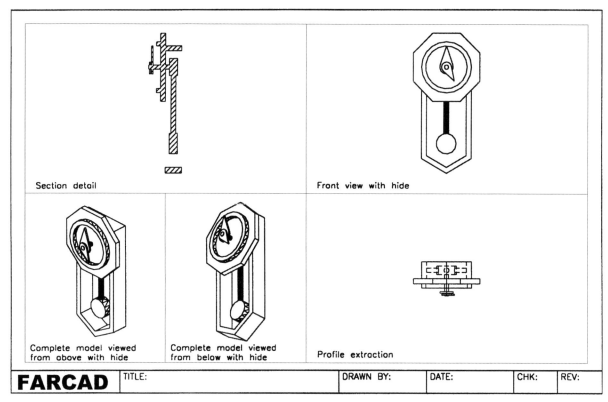

Figure 42.4 Assembled wall clock with profile and section extraction.

External references with solid models

To demonstrate how solid models can be used as wblocks and external references we will use our assembled clock model and add the twelve 'blips' representing 5 minute intervals, so:

1 Close any existing drawing and open your A3SOL sheet with the model tab active

2 With layer MODEL current and UCS BASE, create a solid box with:
   a) corner: 0,0,0
   b) length: 2; width: −1; height: 5

3 Zoom 'tightly' in on this box

4 At the command line enter **BASE <R>** and:
   prompt   Enter base point
   enter   **1,0,0 <R>**

5 Save the model at this stage as **MODR2004\BLIP**

6 Close the existing drawing and open the assembly CLOCK drawing with the model tab active

7 **The following two items are important**:
   a) WCS is current
   b) the lowest front vertex of the clock body is at the point (10,10,0). If it is not, then move the complete clock model so that this lowest vertex is at the stated point.

8 Menu bar with **Insert-External Reference** and:

 prompt `Select Reference File dialogue box`

 respond 1. scroll and pick C:\MODR2004 or named your folder

     2. scroll and pick BLIP

     3. pick Open

 prompt `External Reference dialogue box`

 with   `Name: BLIP`

 respond 1. ensure all on-screen prompts active (tick)

     2. ensure reference type: Attachment

     3. pick OK

 prompt `Attach Xref "BLIP": C:\path name\BLIP.dwg`

     `"BLIP" loaded`

 prompt `Specify insertion point` and enter: **10,−1,150 <R>**

 prompt `Enter X scale factor` and enter: **1 <R>**

 prompt `Enter Y scale factor` and enter: **1 <R>**

 prompt `Specify rotation angle` and enter: **0 <R>**

9 Menu bar with **Modify-3D Array** and:

 prompt `Select objects`

 respond **pick the inserted BLIP object then right-click**

 prompt `Enter type of array`

 enter   **P <R>**

 prompt `Enter the number of items in the array`

 enter   **12 <R>**

 prompt `Specify the angle to fill`

 enter   **360 <R>**

 prompt `Rotate arrayed items as copied`

 enter   **Y <R>**

 prompt `Specify center point`

 enter   **10,−5,125 <R>**

 prompt `Specify second point on axis of rotation`

 enter   **10,15,125 <R>**

10 The inserted BLIP external reference will be arrayed around the clock dial

11 Save the clock drawing at this stage as **MODR2004\BLIPDRG-1**.

Modifying the xref

1 Open MODR2004\BLIP, the original external reference drawing

2 Zoom-in on the object then erase it

3 With layer MODEL current, UCS BASE, model tab active, create a cylinder with:

 a) centre: 1,2.5,0

 b) radius: 2.5

 c) centre of other end option: @0,−1

4 Menu bar with File-Save to automatically update MODR2004\BLIP

5 Close the existing drawing then open MODR2004\BLIPDRG-1

6 Now open the saved MODR2004\BLIPDRG-1 drawing and:

 prompt `Resolve Xref"BLIP": C:\pathname\BLIP.dwg`

     `"BLIP" loaded`

     `"BLIP" reference file may have changed since host`

     `drawing was last saved`

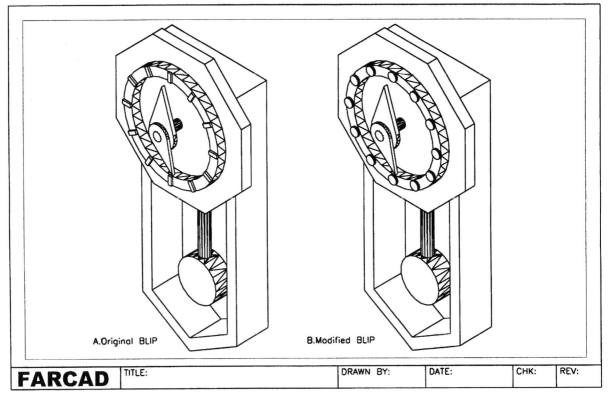

Figure 42.5 External reference example.

7 The clock model should be displayed with the modified cylinder BLIP. Figure 42.5 displays the original and modified xref models.

8 The drawing can now be saved if required

9 *Note*: the bind operations previously investigated in Chapter 23 can be applied to solid model xrefs. You can investigate these in your own time.

This exercise completes the chapter.

Summary

Blocks, wblocks and external references can be used with solid modelling as with any other type of AutoCAD object.

Assignment

In Activity 22, the garden block of MACFARAMUS was created. This object was used for many building projects in various parts of the city of CADOPOLIS. Part of a garden project was unearthed which contained a wall, gate and path. The original garden block has to be saved as a block (no pun intended) and used to create the wall, gate and path.

Activity 23: Using the garden block of MACFARAMUS
1 Open the Activity 21 drawing and create a block of the garden block with:
   a) name: GARDBL
   b) base point: pick the suit – suggest 0,0
   c) select the object
   d) ensure insert units are millimetres

2 *Wall*: with UCS RIGHT
   a) insert the GARDBL at a suitable point
   b) the scales are full size
   c) create a suitable wall layout

3 *Gate*: with UCS RIGHT
   a) insert the GARDBL at the correct point relative to the wall
   b) scales are X: 2, Y: 3, Z: 0.5

4 *Path*: with UCS BASE
   a) insert the GARDBL at the correct point relative to the gate
   b) scales are X: 2, Y: 1, Z: 0.25

5 Optimise your layout and save.

Chapter 43

The setup commands

The setup commands (Drawing, View and Profile) allow drawing layouts to be created by the user. As the profile command has been discussed in a previous chapter, we will only consider the View and Drawing commands in this chapter. These two commands can be summarised as:

View: creates floating viewports using orthographic projection to layout multi and sectional views of 3D solid models

*Draw*ing: generates profiles and sections in viewports which have been created with the Setup View command.

Basically the two commands allow the user to create multiple viewport configurations which will display top, front, end and auxiliary views as well as extracting profile and sections of the model. In other words the two commands will create the same type of layout that has been achieved with our A3SOL template file, the 3D viewpoint command and with the SECTION and PROFILE commands.

We will demonstrate the two commands using previously created models.

Example 1 – the backing plate

1   Open the drawing file MODR2004\BACKPLT from Chapter 32 and:
    a) MVLAY1 tab active
    b) restore UCS BASE with layer MODEL current
    c) in paper space erase the viewports except the 3D view
    d) scale the 3D viewport about the lower left corner by 0.5
    e) in paper space ensure the paper space icon is relative to World by command line entry **UCS <R>** then **W <R>**
    f) in model space, zoom-extents

2   Make a new layer, VPORTS, colour to suit and current

3   Refer to Fig. 43.1.

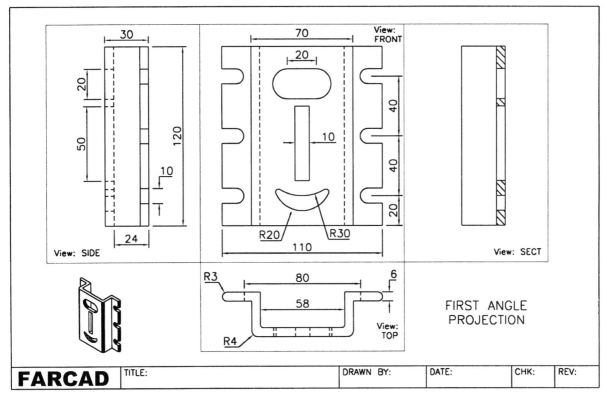

Figure 43.1 The Setup View and Drawing Example 1 – the backing plate.

The Setup View command

1 Select the SETUP VIEW icon from the Solids toolbar and:

prompt	Enter an option [Ucs/Ortho/Auxiliary/Section]
enter	**U <R>** – the Ucs option
prompt	Enter an option [Named/World/?/Current]<Current>
enter	**N <R>** – the named option
prompt	Enter name of UCS to restore
enter	**BASE <R>**
prompt	Enter view scale<1>
enter	**1 <R>**
and	paper space 'entered' with paper space icon displayed
prompt	Specify view center
enter	**200,50 <R>**
and	UCS icon positioned at the entered point
prompt	Specify view center, i.e. reposition the view center?
respond	**<RETURN>**, i.e. accept entered value
prompt	Specify first corner of viewport
enter	**130,25 <R>**
prompt	Specify opposite corner of viewport
enter	**270,85 <R>**
prompt	Enter view name
enter	**TOP <R>**
and	model space returned
prompt	Enter an option [Ucs/Ortho/Auxiliary/Section], i.e. any more options
enter	**X <R>** – to end the command at this stage *and* a top view of the model is displayed in a viewport

2 Menu bar with **Draw-Solids-Setup-View** and:
   prompt   Enter an option [Ucs/Ortho/Auxiliary/Section]
   enter   **O <R>** – the ortho option
   and   paper space 'entered' – note icon
   prompt   Specify side of viewport to project
   respond pick bottom horizontal line of the new viewport and note the midpoint
             snap effect
   prompt   Specify view center
   enter   **200,170 <R>**
   prompt   Specify view center and <RETURN>
   prompt   Specify first corner of viewport
   enter   **130,85 <R>**
   prompt   Specify opposite corner of viewport
   enter   **270,245 <R>**
   prompt   Enter view name
   enter   **FRONT <R>**
   prompt   Enter an option [Ucs/Ortho/Auxiliary/Section]
   enter   **X <R>** to end the command

3 A front view of the model is displayed in a viewport

4 At the command line enter **SOLVIEW <R>** and:
   prompt   Enter an option [Ucs/Ortho/Auxiliary/Section]
   enter   **O <R>** – the ortho option
   prompt   Specify side of viewport to project
   respond pick right vertical line of the new (front) viewport
   prompt   Specify view center
   enter   **80,165 <R>**
   prompt   Specify view center and <RETURN>
   prompt   Specify first corner of viewport
   enter   **130,85 <R>**
   prompt   Specify opposite corner of viewport
   enter   **25,245 <R>**
   prompt   Enter view name
   enter   **SIDE <R>**
   prompt   Enter an option [Ucs/Ortho/Auxiliary/Section]
   enter   **X <R>** to end the command

5 A view of the model from the right is displayed

6 Make the new FRONT viewport active

7 Activate the Setup View command and:
   prompt   Enter an option [Ucs/Ortho/Auxiliary/Section]
   enter   **S <R>** – the section option
   prompt   Specify first point of cutting plane
   enter   **0,0,0 <R>**
   prompt   Specify second point of cutting plane
   enter   **0,120 <R>**
   and   a dashed 'section' line is displayed
   prompt   Specify side to view from
   respond **pick any point to left of section line**
   prompt   Enter view scale
   enter   **1 <R>**
   and   paper space 'entered'. Note the orientation of the paper space icon.
             This is important

prompt	Specify view center
enter	**@0,−125 <R>**
prompt	Specify view center and <RETURN>
prompt	Specify first corner of viewport
enter	**270,85 <R>**
prompt	Specify opposite corner of viewport
enter	**370,245 <R>**
prompt	Enter view name
enter	**SECT <R>**
prompt	Enter an option [Ucs/Ortho/Auxiliary/Section]
enter	**X <R>** to end the command

8 A left view of the model is displayed, but not a section view?

9 *Note*:
   a) In the example I have given co-ordinates for the view centre point, the viewport corners, the section line, etc. It is usual to 'pick these points' on the screen rather than enter co-ordinates. The only reason for the co-ordinate entry was that this was our first example and it should help the user understand the responses to the various prompts.
   b) The view names can be entered as V1, V2, etc. or as TOP, FRONT. This is dependent on the user. Just ensure that you know what view has been allocated to the name.
   c) The Setup View command can be used continually, i.e. when a view has been created, the command line returns the original 'Enter and option' prompt. The user can continue with other options. I ended the command after each new view was created, simply to show how the command can be activated by different methods.
   d) When a view name has been used, it cannot be used again, even if the 'associated' viewport is erased.

Investigating the layers

1 Menu bar with **Format-Layer** and note the new layer names:
   a) FRONT-DIM, FRONT-HID, FRONT-VIS
   b) SECT-DIM, SECT-HAT, SECT-HID, SECT-VIS
   c) SIDE-DIM, SIDE-HID, SECT-VIS
   d) TOP-DIM, TOP-HID, TOP-VIS
   e) VPORTS

2 Each new viewport has a dimension, hidden and visible layer and the section viewport has a hatch layer

3 In model space make the TOP viewport active and activate the Layer Properties Manager dialogue box. The three FRONT, four SECT and three SIDE layers are all:
   a) frozen in this current viewport
   b) frozen in new viewports

4 Using the Layer Properties Manager dialogue box, load the HIDDEN linetype and set the four -HID layers to this linetype. The reason for this will become obvious.

5 The new layers are automatically created with the Setup View command and are viewport specific as follows:
   a) -DIM: for dimensions
   b) -HID: for hidden detail
   c) -VIS: for visible lines
   d) -HAT: for hatching detail if the section option has been used.

The Setup Draw command

1 In model space with any viewport active, set the following system variables from the command line:

command line	*enter*
HPNAME	ANSI31
HPANG	0
HPSCALE	1

2 Select the SETUP DRAW icon from the Solids toolbar and:

and   paper space entered
prompt Select viewports to draw
then   Select objects
respond **pick the four new viewports borders then right-click**

3 The four new viewports will display the model:
   a) with hidden line removal: top, front, right
   b) as a section: left – with no hidden detail

4 What we have achieved is a First Angle Projection layout for the created solid model

5 You should now understand why the HIDDEN linetype was loaded and set to certain layers, and why the three HP system variables were set to the values in step 1. It should be noted that the default hatch pattern is ANGLE

6 Optimise the LTSCALE system variable for the hidden line detail

7 *Note*:
   a) The three command line entries of HPNAME, HPANG and HPSCALE have set the hatch pattern name, angle and scale. This could have been achieved after the hatching was added to the section view with the Modify-Hatch command.
   b) Note the model space icon in each viewport has the same orientation. This is because when new viewports are created, the UCSVP system variable defaults to 1, i.e. the UCS icon will have the same configuration in each new viewport. We will leave it as it is.
   c) The procedure used in this example is the 'basis' for creating an orthographic layout

8 *Task*
   Add dimensions to the new viewports remembering:
   a) the -DIM layers are viewport specific but need to be current in the appropriate viewport
   b) is a named UCS required for this?
   c) Figure 43.1 displays the complete layout with dimensions
      These dimensions have been added without altering the UCS orientation. They are also viewport specific dimensions and not paper space dimensions. Thus this is another method of adding dimensions to a drawing layout
   d) save the model with your own entered file name.

Example 2 – the slip block

In this example we will create another first angle layout from a solid model and add an auxiliary view. We will also create the new viewports by selecting points on the screen rather than by co-ordinate entry. In the exercise, I have omitted the prompts, and given the steps as a series of instructions so:

1 Open model MODR2004\SLIPBL from Chapter 37 and:
   a) in paper space erase the four viewports (MVLAY1 tab)
   b) make a new layer named VPORTS and current
   c) refer to Fig. 43.2

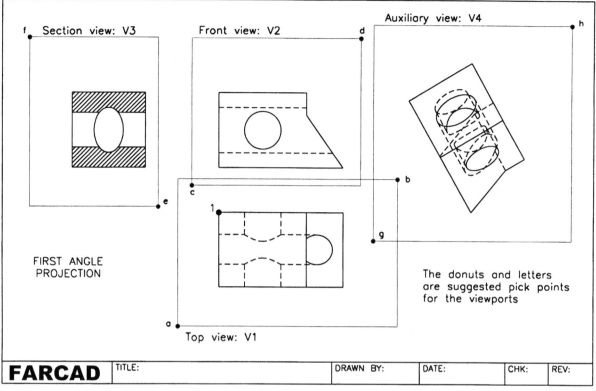

Figure 43.2 The Setup View and Drawing Example 2 – the slip block.

2 Use the Setup View command with:
  a) options: U – the UCS option
  b) options: N – the Named UCS option
  c) enter: BASE – the UCS name
  d) view scale: 0.5
  e) view centre: 190,90 then another <RETURN>
  f) first corner of viewport: pick any pt a (see Fig. 43.2)
  g) opposite corner of viewport: pick any pt b
  h) view name: V1
  i) options: X to end command

3 Use the SOLVIEW command with:
  a) options: U
  b) options: N
  c) name: FRONT
  d) view scale: 0.5
  e) view centre: 190,170 then another <RETURN>
  f) first corner of viewport: pick any pt c
  g) opposite corner of viewport: pick any pt d
  h) view name: V2
  i) options: X

4 In model space make the new viewport active (it should be) then return to paper
  space

5 With the SOLVIEW command use the following:
   a) options: S – section
   b) first point of cutting plane: 50,0
   c) second point of cutting plane: 50,100
   d) side to view from: pick to right of the section line
   e) view scale: 0.5
   f) view centre: 0,100 then another <RETURN>
   g) note the paper space icon orientation
   h) first corner of viewport: pick any pt e
   i) opposite corner of viewport: pick any pt f
   j) view name: V3
   k) options: X

6 Activate the Setup View command and:
   prompt   Enter an option [Ucs/Ortho/Auxiliary/Section]
   enter   **A <R>** – the auxiliary option
   prompt   Specify first point of inclined plane
   respond   **make first new viewport active**
   and   **ENDpoint icon and pick pt1**
   prompt   Specify second point of inclined plane
   enter   **@100<−60 <R>**
   prompt   Side to view from
   respond   **pick 'below' the inclined plane**
   and   note the paper space icon orientation
   prompt   View centre
   enter   **0,175 <R>** then another <RETURN>
   prompt   Specify first corner of viewport and pick any pt g
   prompt   Specify opposite corner of viewport and enter pick any pt h
   prompt   View name and enter: **V4 <R>**
   prompt   Enter an option [Ucs/Ortho/Auxiliary/Section] and enter: **X <R>**

7 Select the Setup Drawing icon and pick the four new viewport borders and the display
   will be:
   a) as a section: the left viewport
   b) with hidden line removal: the other three viewports

8 The user will probably have to load the linetype HIDDEN and 'set' this linetype to the
   – HID layers created. This is a bit of an annoyance with the SETUP commands, i.e.
   having to load the HIDDEN linetype. The user could always load the linetype before
   activating the command.

9 Optimise the LTSCALE variable to suit

10 Note that the added hatch effect is the ANGLE hatch pattern. This is the default setting
   as stated earlier.

11 To alter the hatch pattern, enter model space and make the left (section) viewport
   active. Restore UCS RIGHT.

12 Menu bar with **Modify-Object-Hatch** and:
   prompt   Select associative hatch object
   respond   **pick the hatching**
   prompt   Hatch Edit dialogue box
   respond   1. Type: User-defined
              2. Angle: 45
              3. Spacing: 4
              4. pick OK

13 The model layout is now as Fig.43.2 and can be saved

14 Remember that the VPORTS layer can be frozen/turned off for maximum effect

15 *Note*:
The view centre entry with the section and auxiliary options is perpendicular to the paper space icon. This icon is orientated relative to the:
a) section plane
b) auxiliary inclined plane.

Example 3 – the pipe and flange model

In this third demonstration of using the setup commands, we will create a third angle orthographic layout and add both a sectional view and an auxiliary view to the layout.

1 Open the drawing file MODR2004\FLPIP created in Chapter 33 and:
a) make the MVLAY1 tab active
b) paper space and erase three viewports leaving the 3D view
c) make a new current layer: VPORTS
d) load the linetype HIDDEN
e) model space with UCS BASE and set the LTSCALE variable value to a suitable value
f) set the following hatch variables:
HPNAME: ANSI32
HPANG: 0
HPSCALE: 1
g) paper space scale the 3D viewport by 0.5 and move to the top left of the drawing sheet
h) model space and zoom the model in the 3D viewport to a suitable scale
i) refer to Fig. 43.3 for the viewport layouts

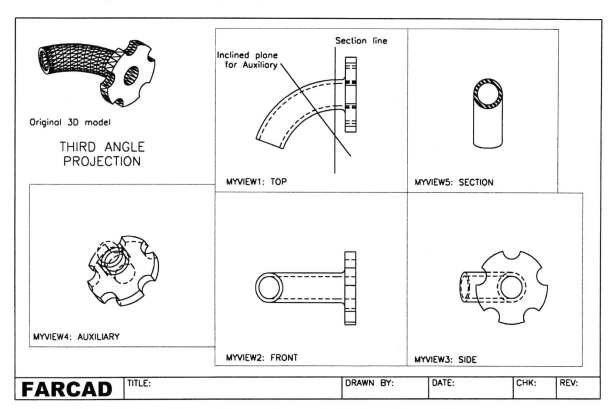

Figure 43.3   The Setup View and Drawing Example 3 – the flange/pipe model.

2 Activate the Setup View command and:
   a) UCS option with UCS BASE
   b) view scale: 0.25
   c) view centre: 200,200
   d) position viewport: pick to suit yourself
   e) name: MYVIEW1
   f) options: X

3 Repeat the Setup View command:
   prompt   Enter an option [Ucs/Ortho/Auxiliary/Section]
   enter   **O <R>** – the ortho option
   prompt   Specify side of viewport to project
   respond **pick lower horizontal line of viewport**
   prompt   Specify view centre
   enter   **200,75 <R>**
   prompt   Specify view centre
   and   **<RETURN>**
   prompt   Specify corners of viewport
   respond **pick to suit your layout**
   prompt   Enter view name
   enter   **MYVIEW2 <R>**
   prompt   Enter an option and continue with next part of exercise

4 Setup View command still active with options:
   a) select the Ortho option
   b) pick right vertical line of second viewport
   c) view centre: pick a point to right to suit
   d) viewport corners: pick points to suit
   e) view name: MYVIEW3
   f) options and enter X <R>

5 SOLVIEW <R> at the command line and:
   a) options: select Auxiliary
   b) inclined plane points: pick to suit – see Fig. 43.3
   c) side to view from: pick 'below' the inclined line
   d) view centre: pick a point to suit
   e) viewport corners: position to suit
   f) view name: MYVIEW4
   g) options
   h) select the section option
   i) cutting plane points: pick points as indicated in Fig. 43.3
   j) side to view from: pick to right of the section line
   k) view scale: 0.25
   l) view centre: pick to suit
   m) viewport corners: pick to suit the layout
   n) name: MYVIEW5
   o) options: X to end command

6 Linetype HIDDEN and three HP variables set?

7 Activate the Setup Drawing command and pick the five viewports to display the layout with hidden line removal and section detail as Fig. 43.3

8 Freeze layer VPORTS if required

9 *Observation*

The three examples should make the user aware of the power of the two setup commands (VIEW and DRAW). From a solid model, the user is able to create first and third angle orthographic layouts and add sectional and auxiliary views as required. Dimensioning layouts obtained with the setup commands is also relatively easy, as specific viewport specific layers are created for this purpose.

I would therefore suggest that once the user knows how to create a composite model from primitives, regions, etc., the only other command that is required is SETUP?

10 Save the exercise as it is now complete.

Summary

1 The set View and Drawing commands allow the user to layout multi-view drawings without the need to create viewports and set viewpoints

2 Both commands can be activated:
   a) by selecting the icon from the Solids toolbar
   b) from the menu bar with Draw-Solids-Setup
   c) from the command line with SOLVIEW and SOLDRAW

3 The Setup View command has options which allow views to be created:
   a) relative to a named UCS
   b) as an orthographic view relative to a selected viewport
   c) as an auxiliary view relative to an inclined plane
   d) as a section view relative to a cutting plane

4 When used, the View command creates viewport specific layers, these being relative to the viewport handle number with the following names:
   -VIS   for visible lines
   -HID   for hidden lines
   -DIM   for dimensions
   -HAT   for hatching but only if the section option is used

5 The View command requires the user to:
   a) enter the view scale
   b) position the viewport centre point
   c) position the actual viewport corners

6 With the Ortho option, both First and Third angle projections can be obtained dependent on which side the new viewport is to be placed

7 The section option requires that the system variables HPNAME, HPANG and HPSCALE are set. It is usual to use the ANSI31 hatch pattern name, but this is not essential. AutoCAD defaults the ANGLE hatch pattern.

8 The Drawing command will display models which have been created with the View command:
   a) with visible and hidden detail
   b) as a section if the section option has been used

9 It is recommended that the linetype HIDDEN be loaded before the Drawing command is used

10 The hidden linetype appearance is controlled by the LTSCALE system variable

11 The user now has two different methods for creating multi-view layouts of solid models:
   a) using the A3SOL template file idea which sets the viewports and viewpoints prior to creating the model. Profiles can then be extracted to display hidden detail.
   b) using the VIEW and DRAWING commands with a solid composite to layout the drawing in First or Third angle projection with sections and auxiliary views as required.
   c) it is now the user's preference as to which method is used

12 Dimensions can now be added to models:
   a) using viewport specific layers
   b) using paper space dimensioning
   c) using the setup commands.

Chapter 44

The final composite

We have now covered virtually every concept of solid modelling within the AutoCAD draughting package. The next chapter will introduce the user to rendering, but before that we will make a final solid model using the various techniques that have been discussed. The model is quite involved, so try not to miss out any of the steps, especially those which set a new UCS position.

The three examples selected to demonstrate the View and Drawing commands in the previous chapter used previously created models. This example will create a new model 'from scratch'.

1 Open your A3SOL template/drawing file with layer MODEL, UCS BASE and make the model tab active. We will use this tab to create the model and then set up our drawing layout with the MVLAY1 tab.

2 Pan the UCS to the lower centre of the screen

3 Set ISOLINES to 6 and refer to Fig. 44.1

4 The new model will be created from five primitives, each requiring a new UCS position.

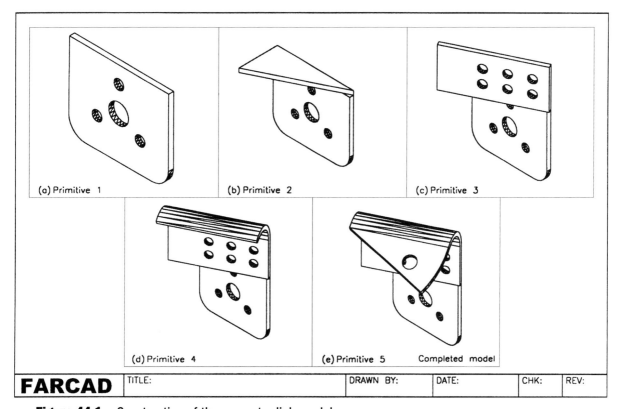

Figure 44.1 Construction of the computer link model.

Primitive 1: the base

1 Rotate the UCS about the X axis by 90 and save as PRIM1

2 Draw a polyline:
   a) *Start point*: 0,50
   b) *Next point*: @60,0
   c) *Next point*: @0,−40
   d) arc option with endpoint: @−10,−10
   e) line option to: @−40,0
   f) arc option with endpoint: @−10,10
   g) line option to: close

3 Zoom-extents then zoom to a scale of 3

4 Solid extrude the polyline for a height of 3 with 0 taper

5 Create two cylinders:
   a) centre: 30,25,0; radius: 6; height: 3
   b) centre: 30,40,0; radius: 3; height: 3

6 Polar array the smaller cylinder about the point 30,25 for 3 items with full circle rotation

7 Subtract the four cylinders from the extruded polyline – Fig. 44.1(a).

Primitive 2: wedge on top of first primitive

1 UCS PRIM1 current

2 Set a new 3 point UCS position with:
   a) origin: 0,50,0
   b) X axis: 60,50,0
   c) Y axis: 0,50,3
   d) save as: PRIM2

3 Create a wedge with:
   a) corner: 0,0,0
   b) length: 60; width: −3; height: −30

4 Rotate 3D this wedge:
   a) about the X axis
   b) with 0,0,0 as a point on the axis
   c) for 90 degrees

5 Union the wedge and the extruded polyline – Fig. 44.1(b).

Primitive 3: box on top of wedge

1 UCS PRIM2 current

2 Set a new 3 point UCS position with:
   a) origin: 60,0,0
   b) X axis: 0,30,0
   c) Y axis: 60,0,−3
   d) save as: PRIM3

3 Create a solid box with
   a) corner: 0,0,0
   b) length: 67.08; width: 30; height: 3. Why 67.08?

4   Create a cylinder with:
    a) centre: 10,10,0
    b) radius: 3 and height: −3

5   Rectangular array the cylinder:
    a) for 2 rows and 3 columns
    b) row offset: 10 and column offset: 15

6   *a*) subtract the six cylinders from the box
    b) union the box and the composite – Fig. 44.1(c).

Primitive 4: curved extension on top of box

1   Pan the model to lower part of screen

2   UCS PRIM3 current

3   Set a new 3 point UCS position with:
    a) origin: 0,30,−3
    b) X axis: 67.08,30,−3
    c) Y axis: 0,30,0
    d) save as: PRIM4

4   Zoom-in on the 'free edge' of the box

5   Draw two line segments with:
    a) *Start point*: 0,0
    b) *Next point*: @0,−15
    c) *Next point*: @50,0

6   Draw a polyline about the 'top rim' of the box using ENDPOINT snap and the close option

7   With the solid revolve command:
    a) objects: enter L <R><R> – to select the polyline
    b) options: enter O <R> – object option
    c) object: pick the left end of long construction line
    d) angle of revolution: enter 120

8   Erase the two line segments

9   Zoom-previous to restore original view

10 Union the revolved component and the composite – Fig. 44.1(d).

Primitive 5: final curved component

1   UCS PRIM4 current

2   Set a new 3 point UCS position with:
    a) origin: 67.08,−22.5,−12.99
    b) X axis: 0,−22.5,−12.99
    c) Y axis: 67.08,−24,−15.59
    d) save as: PRIM5
    e) can you work out the three sets of co-ordinates?

3   Zoom-in on the 'free end' of the curved component

4   Draw a polyline about the free end of the curved component using ENDPOINT snap and the close option

5   With the Solid revolve command:
   a) objects: enter L <R><R> – to select the polyline
   b) options: enter Y <R> – the Y axis
   c) angle: enter −30

6   Create a cylinder with:
   a) centre: 45,0,15
   b) radius: 5
   c) centre of other end: @0,10,0

7   Subtract the cylinder from the revolved component, then union the revolved component with the cylinder – Fig. 44.1(e)

8   Zoom-previous to restore the original view

9   The model is now complete, so:
   a) Gouraud shade and 3D orbit – impressive?
   b) restore 2D wire-frame representation at the original viewpoint

10 *a*) restore UCS BASE
   b) save as MODR2004\COMPLINK.

Laying out the viewports

This part of the exercise will use the MVLAY1 tab with all options of Setup View command.

1   Pick the MVLAY1 tab name

2   In paper space:
   a) erase three viewports but leave the 3D viewport
   b) stretch (crossing option) the vertical right edge of the 3D viewport by @−100,0
   c) in model space, UCS BASE, layer Model current and zoom the model to suit
   d) return to paper space
   e) make a new layer VPORTS, colour to suit and current

3   *a*) load linetype HIDDEN
   b) set the following variables:
      HPNAME: ANSI31; HPANG: 0; HPSCALE: 0.5
   c) set the LTSCALE value to suit which may change after the layout has been created

4   Activate the Setup View command with:
   a) UCS option with BASE
   b) view scale: 1
   c) view centre: 175,75
   d) viewport corners: pick to suit
   e) view name: TOP
   f) options: X

5   Using the SOLVIEW command:
   a) UCS option with PRIM1 as the named UCS
   b) view scale: 1
   c) view centre: 175,200
   d) viewport corners: pick to suit
   e) view name: FRONT
   f) exit command or continue with command

6 SOLVIEW command with:
   a) Ortho option
   b) side: pick right vertical side of the second viewport
   c) view centre: 50,200
   d) viewport corners: pick to suit
   e) view name: SIDE
   f) exit command or continue with command

7 Make the FRONT viewport active and with the Setup View command:
   a) activate the Section option
   b) cutting plane points: 30,0 and 30,120
   c) side to view from: pick a point to left of section line
   d) view scale: 1
   e) view centre: 0,−110
   f) viewport corners: pick to suit
   g) view name: SECT
   h) exit or continue with command

8 The final SOLVIEW command is with the TOP viewport active and:
   a) the Auxiliary option
   b) first point of inclined plane and with the first viewport active, pick ENDpoint of pt1 (see Fig. 44.2)
   c) second point of inclined plane: PERP to line 23
   d) side to view from: pick to left of inclined line
   e) view centre: 0,200
   f) viewport corners: pick to suit
   g) view name: AUX
   h) end the command

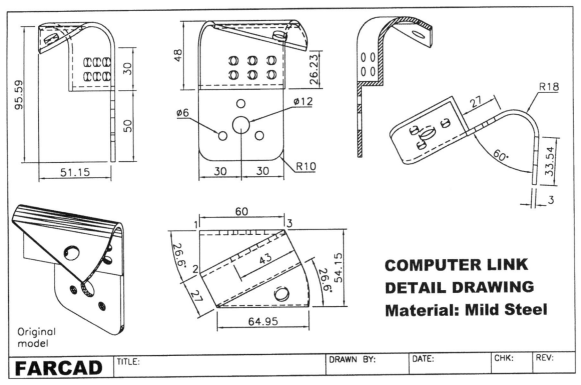

Figure 44.2 Computer link detail drawing using the setup commands.

9   Using the Setup Drawing command, pick the five viewports to display hidden detail and a section view – linetype HIDDEN loaded?

10   Now optimise the LTSCALE system variable if required

11   *Tasks*
   a) Interrogate the model in the 3D viewport:
      Area: 20196.01
      Mass: 26758.64
   b) Using the viewport specific -DIM layers, add the dimensions displayed in Fig. 44.2. The UCS in the new created viewports should be 'set' to allow this as the UCSVP system variable always defaults to 1 when a new viewport is created. Note that a paper space zoom of the viewport being dimensioned will assist with the dimensions.
   c) Freeze the VP and VPORTS layers
   d) In paper space, optimise your drawing with suitable text
   e) Save the completed exercise – worth the effort?

Assignment

Activity 24: Dispenser of MACFARAMUS

One of the discoveries in the city of CADOPOLIS was a container which was thought to be a dispenser belonging to MACFARAMUS. It is this container which has to be created as a solid model and then displayed with the setup commands.

1   Use your template file with the Model tab active to create the model then use the MVLAY1 tab as the chapter example to complete the layout

2   Make two new layers BODY blue and TOP green

3   With UCS FRONT, draw the two shapes using the reference sizes in Fig. 44.3. Use the start points given. Draw as lines/arcs then use the join option of the modify polyline command to convert the segments into single polylines.

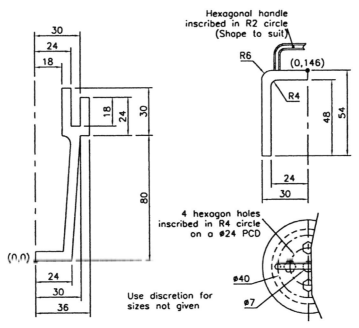

Figure 44.3   Reference details for activity 24.

4   Solid revolve the two polylines for a full circle

5   Create the holes in the top

6   Create a handle from a hexagon, the actual shape being at your discretion – I extruded a hexagon along a polyline path

7   Use the VIEW and DRAWING commands to create a multi-view layout to display:
    a) top and front views with hidden detail
    b) an auxiliary view through an inclined plane at 45 degrees
    c) four section views through vertical cutting planes similar to those shown
    d) a 3D view of the model
    e) *Notes*:
       i) I used a view scale of 0.6
       ii) the actual orthographic layout can be first or third angle projection – it is your choice

8   *Observation*
    This activity should highlight a problem when the setup commands are used with a model containing more than one part. The dispenser has a top and a body, but when the section option is used, the same hatching is added to both parts.

9   *Question*
    How did I achieve the correct hatching using the two setup commands with different hatching added to the two separate parts. Think about using layers!

Chapter 45

Rendering

Rendering is a topic with its own terminology and we will discuss this terminology by rendering two previously created models. The reader should realise that this chapter is only a brief introduction to rendering. As we have created several interesting models, it seems reasonable that we investigate the next step in the modelling process, i.e. the production of rendered images.

What is render?

Rendering is a process which creates an image (usually in colour) of a 3D surface or solid model. This image is created from *a scene using a view with lights*.

How is render activated?

AutoCAD render is automatically loaded into memory when the RENDER command (or any render option) is selected. The RENDER command can be activated with:
a) the menu bar selection **View-Render-Render**
b) the RENDER icon from the Render toolbar
c) entering **RENDER <R>** at the command line

The three methods give the Render dialogue box as Fig. 45.1, which (at present) has three main areas:
1. the rendering types
2. the scene which is to be rendered
3. the rendering destinations

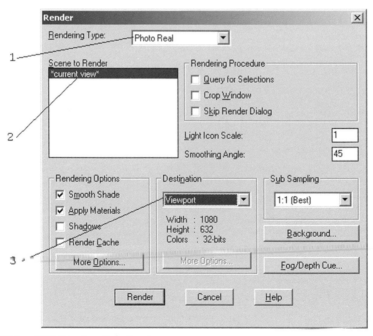

Figure 45.1   The Render dialogue box.

Rendering types

AutoCAD has three 'types' of rendering, these being:
1. Render: the basic AutoCAD render option which allows models to be rendered without the need for scenes, lights or materials. This is the default 'setting'.
2. Photo Real: is a photo-realistic renderer which can display bitmapped and transparent materials. Volumetric and mapped shadows can also be generated.
3. Photo Raytrace: is a photo-realistic raytrace renderer which can generate reflections, refraction and precise shadows

Rendering destination

The rendered image of the model can be:
1. displayed in the current viewport of a multi-screen layout
2. displayed in the render window
3. saved to a file for future recall

Scene to render

Allows the user to select a named scene of the model to render. The user can 'make' several scenes using different views of the model with various lights and materials added. AutoCAD provides a 'default' scene for rendering purposes. This default scene uses a distant light which cannot be modified by the user. At all times it should be remembered that:

*** A scene is a view with lights added ***

The AutoCAD lights

Adding lights to a model layout immediately improves the render appearance and lights can be used to illuminate a complete model or to highlight specific parts of the layout. AutoCAD 2004 has four 'types' of light available, these being ambient, distant, point and spot.

Ambient light

1   Provides a constant illumination to all surfaces of a model

2   It is always 'there' and does not originate from any particular source

3   The user has control of the intensity of the ambient light

4   Generally the ambient light intensity should be a low value or the model will be displayed 'too bright'

5   The default ambient intensity value is 0.3

6   Ambient light on its own does not produce good rendered images.

Distant light

1   Gives a parallel beam in a particular direction

2   The user specifies the target point and the light source location

3   Think of a torch shining at an object

4   Distant light rays extend to infinity on either side of the light source

5   The distant light intensity is **not** affected by the distance of the source from the target

6   It is recommended that distant lights are positioned at the extents of the drawing

7   Distant lights are used to give a 'uniform lighting' facility

8   A single distant light simulates the sun.

Point light

1   A point light emits light in all directions from its position

2   The user specifies the point light location

3   Think of a light bulb

4   The intensity of a point light is affected by the distance from the model

5   Point lights are used for general lighting effects

6   Point lights are used with spot lights for lighting effects.

Spot light

1   Gives a directional cone of light

2   The user specifies the direction of the light and the size of the cone

3   The intensity of a spot light diminishes with the distance from the model

4   Spot lights have 'hot-spots' and 'fall-off angles' that determine how the light diminishes at the edge of the cone

5   Spot lights can be used to highlight specific features on a model.

Point lights, distant lights and spot lights are represented in a drawing with symbols, the light name being displayed within the light symbol – Fig. 45.2

POINT LIGHT   DISTANT LIGHT   SPOT LIGHT
SYMBOL   SYMBOL   SYMBOL

Figure 45.2   Light symbols with 'names' added.

Note

1   Lights are essential for rendering and their position in relation to the model is very important. AutoCAD will position a light in the centre of the active viewport, irrespective of the model position. The user must know exactly where the light is to be positioned relative to the model. This can be achieved using co-ordinates and object snaps. It is also important for the user to know the basic sizes of the model and where the model is situated on the drawing screen. This is one of the main reasons that I use the 0,0,0 origin point when creating models.

2   The basic 'order' with rendering is:
    a) create the model
    b) make a view for a particular model 3D viewpoint
    c) position lights – with or without shadows
    d) add materials to the model parts
    e) make a scene from a view and lights
    f) render with a type and to a destination.

Models to be rendered

Two previously created solid models will be used to demonstrate how lights, scenes, materials, shadows, etc. can be added to a layout and produce a rendered coloured image. The models selected are the extruded backing plate and the wall clock created from wblocks.

Render Example 1 – the solid model backing plate

When render a model, the user will activate several dialogue boxes. It is not my intention to display these dialogue boxes (other than Render) in the exercises. The necessary steps to activate parts of these dialogue boxes will be given, and the user can investigate other options of these dialogue boxes at their leisure.

1 Open your saved drawing MODR2004\BACKPLT from Chapter 32

2 Make the model tab active and set a SE Isometric viewpoint

3 *a*) Ensure UCS BASE and layer Model current
   b) Ensure that the model is positioned with the midpoint of the front bottom edge at the origin. It should be, but if it is not, move the model so that it is. This is import-ant for positioning the lights
   c) Display the Render toolbar

4 For maximum effect we want a shadow effect from the lights. This means that we require objects onto which the shadows 'can be cast' and will therefore create a base for the model to stand on and a back wall. From the menu bar select **Draw-Solids-Box** and create two box primitives with:

	box 1	*box 2*
corner:	−70,−20,0	−70,78,0
length:	200	200
width:	100	2
height:	−2	150
colour no:	41	41

5 Union the two box primitives then zoom-extents. The complete model layout should be displayed

6 *The view*
   As stated earlier, once the model has been created a view should be saved, and we will use our existing screen layout for this, so at the command line enter **VIEW <R>** and:

   prompt   View dialogue box
   respond 1. pick Named View tab
              2. pick New
   prompt   New View dialogue box
   respond 1. View name: enter V1
              2. Current display and Save UCS with view active
              3. UCS name: BASE
              4. pick OK
   prompt   View dialogue box
   with   V1 listed with details
   respond pick OK

7 Note that −VIEW <R> will allow command line entry, the user selecting the save (S) option then entering the view name

8 *The lights*
Select the LIGHTS icon from the render toolbar and:

prompt   Lights dialogue box
respond   1.   set Ambient Light Intensity to 0.3
          2.   scroll at New and pick Point Light
          3.   pick New
prompt   New Point Light dialogue box
respond   1.   Light name: enter P1
          2.   Intensity: alter to 50
          3.   pick Modify
prompt   Drawing screen returned with rubber band effect to light position which is at the 'centre of the viewport', i.e. the screen in this case
and   Enter light location<current>
enter   **0,−50,100 <R>**, i.e. in front of the model
prompt   New Point Light dialogue box
respond   1.   activate Shadows on – tick
          2.   pick Shadows Options
prompt   Shadows Options dialogue box
respond   1.   ensure Shadow Volumes/Ray Traced Shadows active
          2.   pick OK
prompt   New Point Light dialogue box
respond   pick OK
prompt   Lights dialogue box with P1 listed
respond   1.   scroll at New and pick Distant Light
          2.   pick New
prompt   New Distant Light dialogue box
respond   1.   Light name: enter D1
          2.   Intensity: alter to 0.5
          3.   pick Modify
prompt   Drawing screen returned with rubber band effect to light position at centre of screen
and   Enter light direction TO
enter   **50,0,30 <R>**
prompt   Enter light direction FROM
enter   **@20,−30,30 <R>**
prompt   New Distant Light dialogue box
respond   1.   activate Shadows on
          2.   pick Shadows Options
prompt   Shadows Options dialogue box
respond   1.   ensure Shadow Volumes/Ray Traced Shadows active
          2.   pick OK
prompt   Distant Light dialogue box
respond   pick OK
prompt   Lights dialogue box with D1 and P1 listed
respond   pick OK

9 *The scene*
Menu bar with **View-Render-Scene** and:

prompt   Scenes dialogue box
respond   pick New
prompt   New Scenes dialogue box
respond   1.   enter Scene name: SC1
          2.   pick Views: V1
          3.   pick Lights: *ALL*, i.e. both P1 and D1
          4.   pick OK
prompt   Scenes dialogue box with SC1 listed
respond   pick OK

10 *The first render*
    At the command line enter **RENDER <R>** and:
    prompt   Render dialogue box
    respond 1. Rendering Type: scroll and pick Photo Real
               2. Scene to Render: pick SC1
               3. Destination: scroll and pick Render Window
               4. Rendering Options: ensure Smooth Shade, Apply Materials and Shadows
                  are active (tick)
               5. pick Render
    prompt   Render window will be displayed with a coloured image of
               the backing plate
    and   The rendered image should be reasonably impressive and the user should
               observe:
               a) two shadow effects
               b) a white background
               c) the light effect on the model being 'too bright'?
    respond pick AutoCAD from the Windows taskbar to return to the drawing
               screen

11 The user has now to decide whether the lights are in the correct position, is the inten-
    sity set correctly, etc. We will leave the light settings as they are

12 *The background*
    We now want to add a background to enhance the display and AutoCAD allows ren-
    dered images to be displayed with four 'types' of background, these being:
    a) the default white background
    b) a one coloured background
    c) a gradient background of three colours
    d) a 'picture' background of an already saved image

13 Select the BACKGROUND icon from the Render toolbar and:
    prompt   Background dialogue box
    respond 1. Select Gradient
               2. Accept the RGB colour settings
               3. Alter Horizon: 0.7; Height: 0.5; Rotation: −20
               4. pick Preview then OK

14 Render scene SC1 with Photo Real to the Render Window and the model image
    will be displayed with the set gradient background. This is the second render of the
    model.

15 Return to the AutoCAD screen

16 *Attaching a material*
    The rendered image of the model is displayed with the colours of the primitives from
    which it was created. AutoCAD has a library which allows the user to attach different
    materials to surface and solid models, so select the Materials Library icon from the
    Render toolbar and:
    prompt   Materials Library dialogue box
    with   *a*) Materials in Current Drawing – Global
               b) Current Library list
    respond 1. scroll at Current Library list
               2. pick WOOD-WHITE ASH
               3. scroll at Preview, pick Cube then pick Preview
               4. pick <-Import
    and   WOOD-WHITE ASH added to Current Drawing list
    respond pick OK

17 *Note*:
1. The current library list will display either a few material names or a large number of names. If only a few names are displayed then it will be necessary to open an **MLI** file. This is achieved by:
   a) picking Open from the Materials Library dialogue box
   b) scrolling and opening the render.mli files from Support
2. Although WHITE ASH has been selected, any material can be used for this exercise

18 Menu bar with **View-Render-Materials** and:
prompt   Materials dialogue box
respond 1. pick WOOD-WHITE ASH
            2. pick Attach<
prompt   Select objects to attach "WOOD-WHITE ASH" to:
respond pick the composite and right-click
prompt   Materials dialogue box
respond pick OK

19 Now render scene SC1 to the render window to display the model with a material and a background. The result should be quite impressive? This is the third render.

20 The rendered display is a bitmap image and can be saved for future recall into AutoCAD or to other graphics type packages, so from the render menu bar select **File-Save** and:
prompt   Save BMP dialogue box
respond 1. named folder should be current
            2. enter file name: **BACKPLT**
            3. pick OK

21 Return to AutoCAD and save the screen layout with a suitable name. This will save the model and the lights, scene, materials, etc.

22 A screen dump of the model is displayed in Fig. 45.3

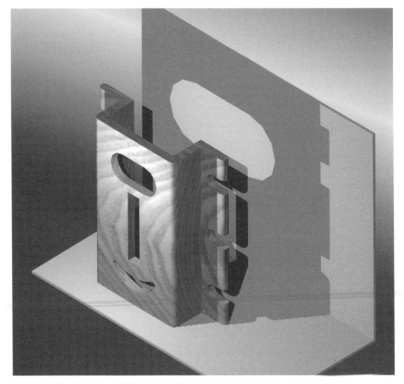

Figure 45.3   The rendered Backing Plate with materials attached.

23 This first exercise is now complete

24 *Note*:
  a) During the render process, we used the render command three times:
  i) when lights had been added
  ii) when the background was set
  iii) when the material was attached to the model
  b) I would suggest to the user that until they become proficient with the render process, that this procedure be adopted to obtain a rendered image.
  c) The various 'operations' to position lights, set a background, add materials, etc. can all be achieved with a single render.

Render Example 2 – the wall clock

Preparing a model for rendering follows the basic procedure used with Example 1, i.e. save a view, position lights, create a scene, attach materials, render the scene. For our second example we will use the wall clock model and attach several materials to the various parts of the model.

1 Open model CLOCK of the wall clock created in Chapter 41

2 Make the model tab active

3 Restore UCS BASE and check that the front lower vertex of the clock body is positioned at 0,0. If it is not, then move it to this position

4 Create a box primitive with:
  a) corner: $-90,40,-50$
  b) length: 180; width: 2; height: 240
  c) colour: number 201

5 Now zoom-extents and at the command line enter **– VIEW <R>** and:
  prompt Enter an option
  enter   **S <R>** – the save option
  prompt Enter view name to save
  enter   **V1 <R>**

6 Position the following three lights:
  1. Point light with:
    a) name: P1
    b) intensity: 100
    c) location: 0,20,300, i.e. modify
    d) no shadow effect
  2. Point light with:
    a) name: P2
    b) intensity: 50
    c) location: 0,−90,150
    d) shadows on: shadow volumes/ray trace shadows
  3. Distant light with:
    a) name: D1
    b) intensity: 0.75
    c) direction TO: −50,−20,150
    d) direction FROM: @ −50,−50,50
    e) shadows (volumes/ray traced) on

7 Make a scene named SC1 with view V1 and all three lights

8 Using the materials library icon, import the materials COPPER, MARBLE GREEN, PINK MARBLE, WOOD-WHITE ASH, BLUE PLASTIC and any other material of your choice

9 With the Materials icon, attach the imported materials to the following parts of the model:
   a) COPPER: pendulum, spindle and mechanism
   b) MARBLE GREEN: octagonal face
   c) PINK MARBLE: circular dial
   d) WOOD-WHITE ASH: body
   e) BLUE PLASTIC: the hands

10 Set a gradient background of your choice

11 Render scene SC1 with:
   a) photo real
   b) render window
   c) shadow options on

12 Save the rendered image to your named folder and save the AutoCAD drawing as it has the lights, materials, etc. set

13 The rendered image of the clock is displayed in Fig. 45.4

14 Note that I have rendered the clock model with the blips

15 This second exercise is now complete.

Figure 45.4 The rendered Wall Clock with materials attached.

3D orbit with materials attached

AutoCAD 2004 allows the 3D orbit command to be used with materials attached. If you have tried the 3D orbit command, the materials may not have been 'retained' during the real-time rotation. To achieve the required effect:

a) a good quality, high memory graphics card is required

b) certain additional settings must be activated

To demonstrate the effect:

1 Open the saved backing plate model which has lights and a wood material attached

2 Activate the model tab

3 Menu bar with **Tools-Options** and:
   prompt   Options dialogue box
   respond 1. refer to Fig. 45.5
              2. pick the System tab
   prompt   System selection
   respond pick Properties at the Current 3D Graphics Display
   prompt   3D Graphics System Configuration dialogue box
   respond 1. Render options active
              2. Enable lights and materials active
              3. Enable textures with High Quality/Slower active
              4. Acceleration: Hardware active
              5. Select driver: scroll and select **wopengl8.hdi** (if available)
   then   pick Apply & Close
   prompt   Options dialogue box
   respond pick OK

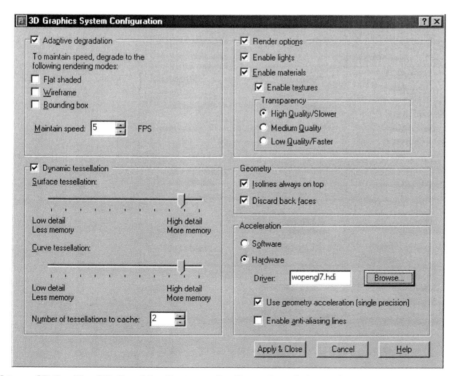

Figure 45.5   The 3D Graphics Systems Configuration dialogue box.

4   Now:
    a) render the model with the Viewport as the destination
    b) activate the 3D orbit command
    c) materials displayed with rotation?
    d) if not, you may require a better graphics card, although I **WOULD NOT RECOM-MEND** that you rush out and buy one!

This chapter has introduced the reader to how models can be rendered. Hopefully you will investigate the topic in more detail as rendering is a fascinating subject and the results really enhance the final appearance of models.

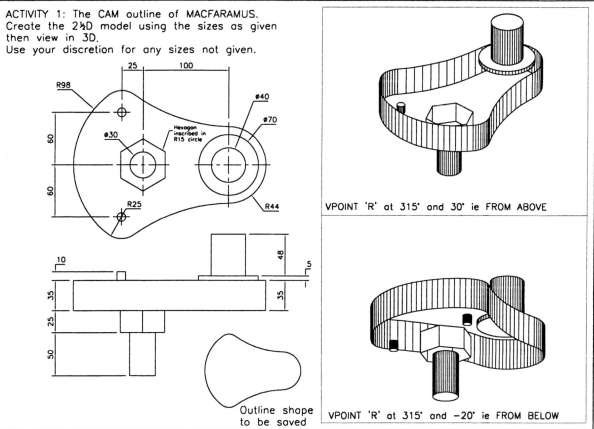

ACTIVITY 1: The CAM outline of MACFARAMUS.
Create the 2½D model using the sizes as given
then view in 3D.
Use your discretion for any sizes not given.

R98

25   100

ø40

ø70

Hexagon
inscribed in
R15 circle

ø30

60

60

R25

R44

10

48

5

35

35

25

50

Outline shape
to be saved

VPOINT 'R' at 315° and 30° ie FROM ABOVE

VPOINT 'R' at 315° and -20° ie FROM BELOW

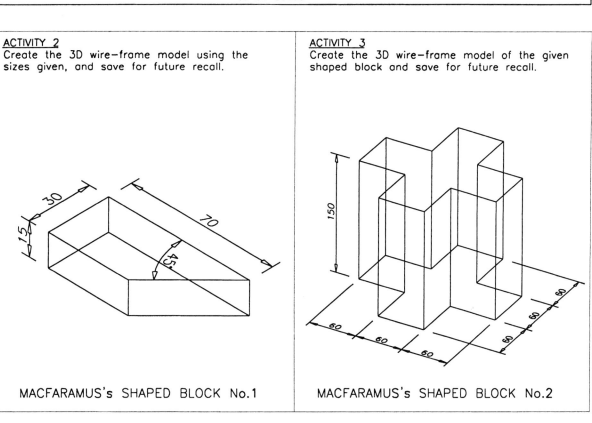

ACTIVITY 2
Create the 3D wire-frame model using the
sizes given, and save for future recall.

30

15

70

45°

MACFARAMUS's SHAPED BLOCK No.1

ACTIVITY 3
Create the 3D wire-frame model of the given
shaped block and save for future recall.

150

60

60

60

60

60

60

MACFARAMUS's SHAPED BLOCK No.2

ACTIVITY 4
Create the 3D wire—frame model of the given
shaped block and save for future recall.

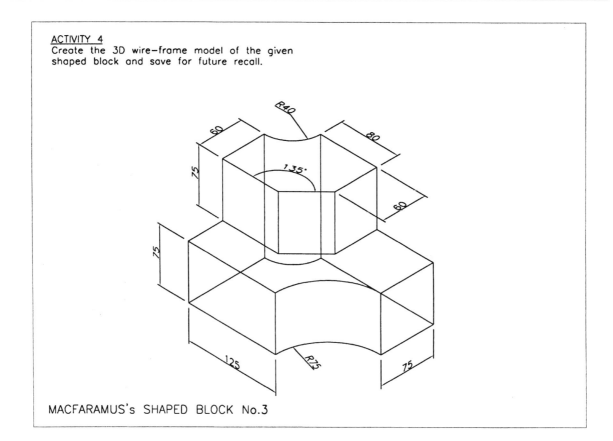

MACFARAMUS's SHAPED BLOCK No.3

ACTIVITY 5
Dimension the shaped wire frame block from activity 2.
Save for future recall.

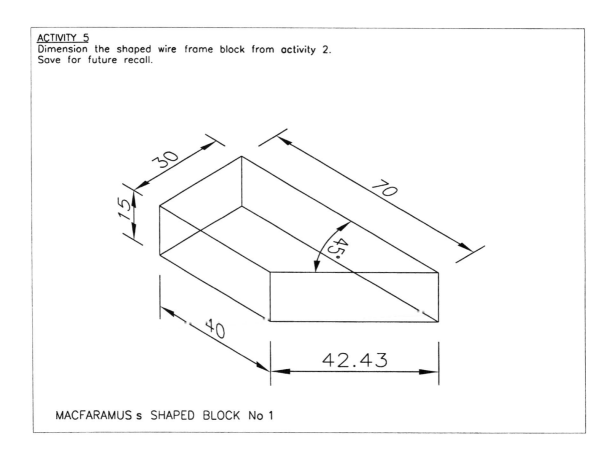

MACFARAMUS s SHAPED BLOCK No 1

ACTIVITY 6

a) Create the wire-frame model using the sizes given
b) Set and save a UCS for each plane on the model. Four UCS names and orientations are suggested
c) Dimension the model
d) Save for future recal
e) Assist MACFARAMUS with his required calculations.

Suggested UCS names and orientations

SLOPE1

SLOPE2

SLOPE3

SLOPE4

MACFARAMUS's RECTANGULAR TOPPED PYRAMID

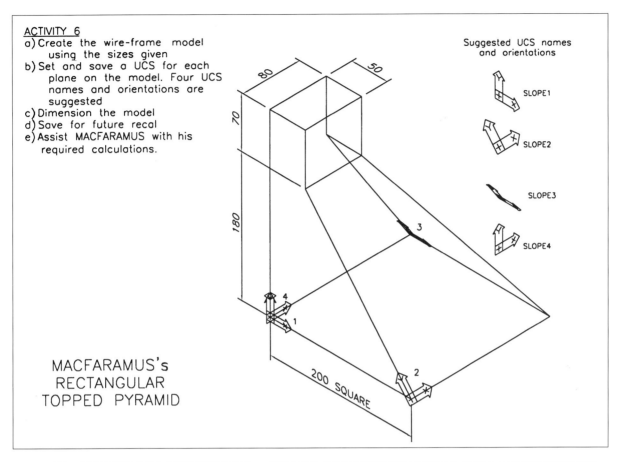

ACTIVITY 7: MACFARAMUS's RECTANGULAR TOPPED PYRAMID

Using the slope and vertical saved UCSs, add hatching to the 'planes' using the hatch information given.

Surfaces	Pattern	Angle, scale
Four slopes	BRICK	2, 0
Four verticals	BRSTONE	1, 0
One top	EARTH	1.5, 0

SE Isometric 3D View

NE Isometric 3D View

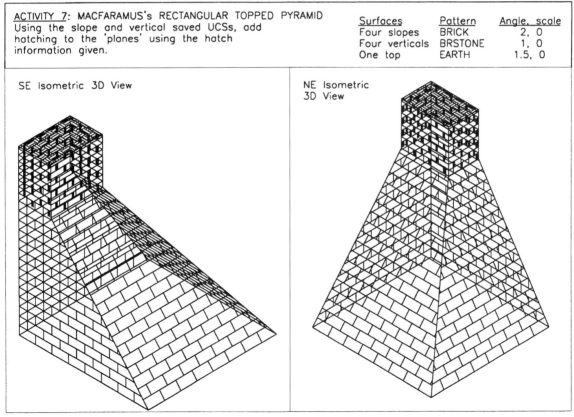

ACTIVITY 8: MACFARAMUS's RECTANGULAR TOPPED PYRAMID
Set the four viewport configuration as shown.

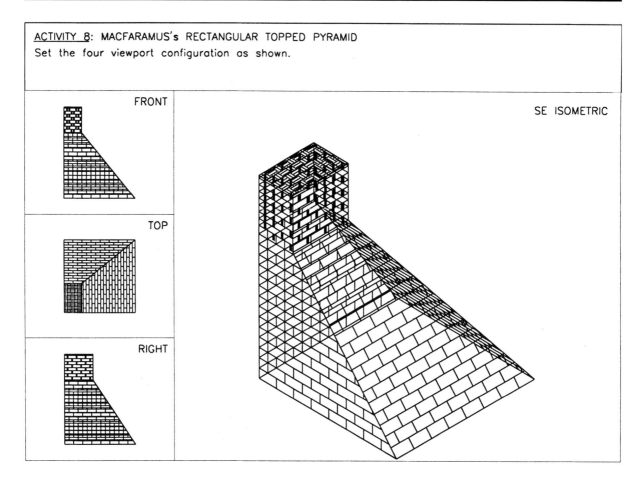

FRONT

TOP

RIGHT

SE ISOMETRIC

ACTIVITY 9
1. Create a wire-frame model using the sizes given
2. Use the 3DFACE command to covert the wire-frame model into a surface model
3. Polar array as instructed

HAXAGONAL COLUMN of MACFARAMUS

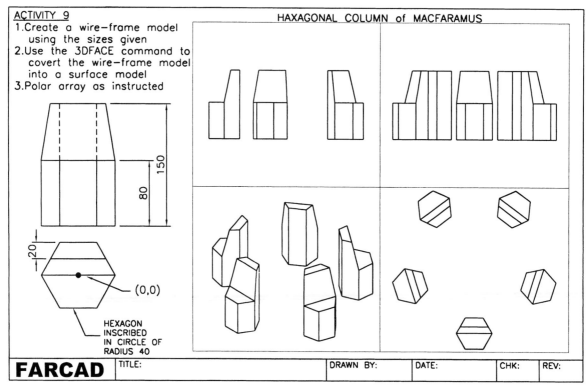

150

80

20

(0,0)

HEXAGON
INSCRIBED
IN CIRCLE OF
RADIUS 40

FARCAD

TITLE:		DRAWN BY:	DATE:	CHK:	REV:

ACTIVITY 10
Using the reference sizes create a 3D wire-frame model then
add ruled surfaces using layers correctly.

ORNAMENTAL FLOWERBED
of MACFARAMUS

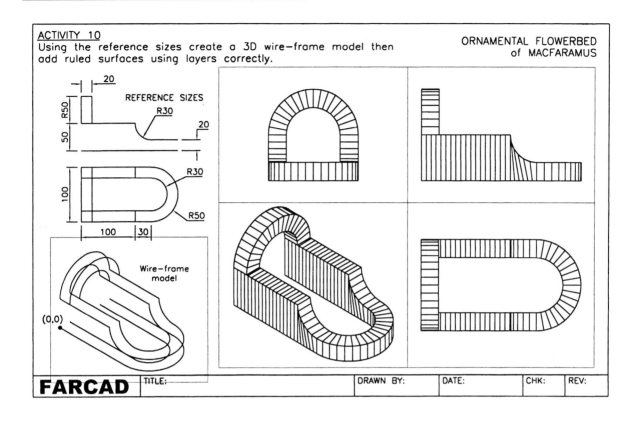

REFERENCE SIZES

R30

R30

R50

Wire-frame
model

(0,0)

FARCAD | TITLE: | | DRAWN BY: | DATE: | CHK: | REV:

ACTIVITY 11
Using the reference sizes as
a guide, design the two path
curve profiles and generate
revolved surface models of
the table and seat.
Use discretion for design.

GARDEN FURNITURE ARRANGEMENT of MACFARAMUS

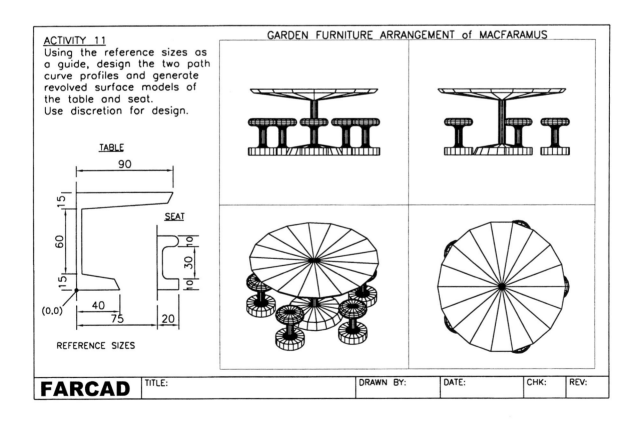

TABLE

SEAT

(0,0)

REFERENCE SIZES

FARCAD | TITLE: | | DRAWN BY: | DATE: | CHK: | REV:

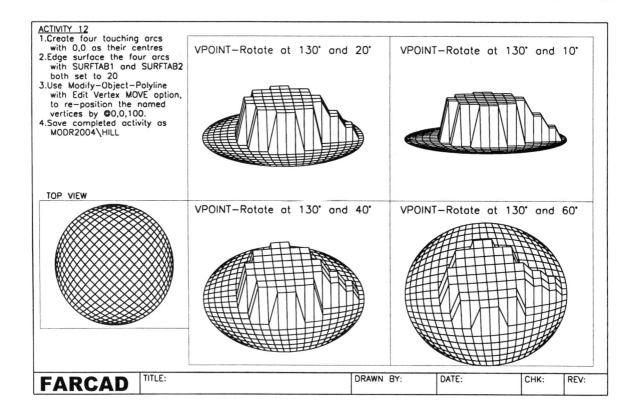

ACTIVITY 12
1. Create four touching arcs with 0,0 as their centres
2. Edge surface the four arcs with SURFTAB1 and SURFTAB2 both set to 20
3. Use Modify−Object−Polyline with Edit Vertex MOVE option, to re−position the named vertices by ⌀0,0,100.
4. Save completed activity as MODR2004\HILL

TOP VIEW

VPOINT−Rotate at 130° and 20°

VPOINT−Rotate at 130° and 10°

VPOINT−Rotate at 130° and 40°

VPOINT−Rotate at 130° and 60°

FARCAD	TITLE:		DRAWN BY:	DATE:	CHK:	REV:

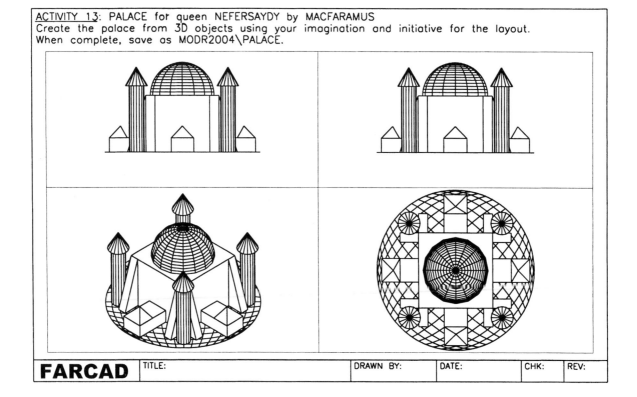

ACTIVITY 13: PALACE for queen NEFERSAYDY by MACFARAMUS
Create the palace from 3D objects using your imagination and initiative for the layout.
When complete, save as MODR2004\PALACE.

FARCAD	TITLE:		DRAWN BY:	DATE:	CHK:	REV:

ACTIVITY 14
1. Recall the drawing layout MODR2004\CHESS
2. Design the other chess pieces
3. Save the completed layout for rendering.

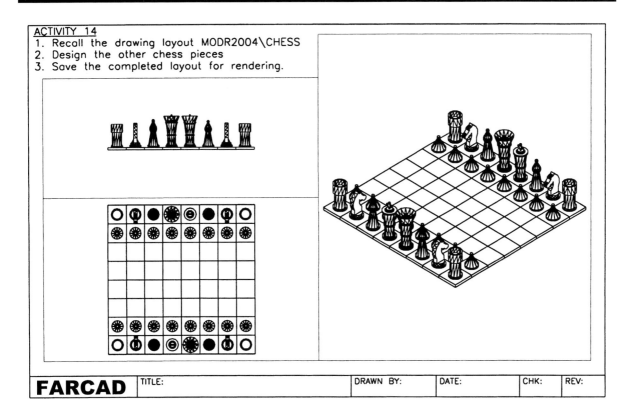

FARCAD	TITLE:		DRAWN BY:	DATE:	CHK:	REV:

ACTIVITY 15
1. Open the saved edge surface layout MODR2004/HILL
2. Insert the MODR2002 PALACE model, centred on the top of the hill
3. Create viewports to display the complete model 'to perfection'

PALACE of Queen NEFERSAYDY
by
MACFARAMUS

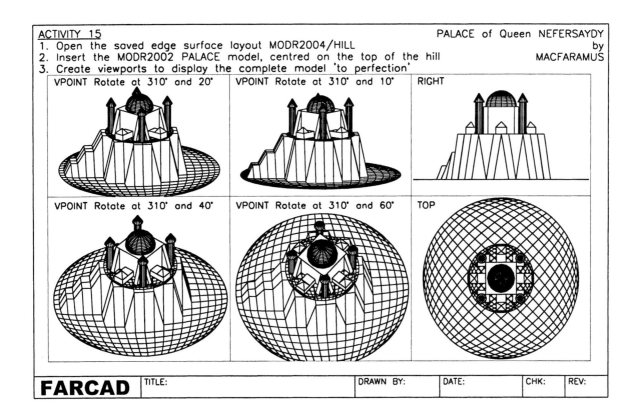

VPOINT Rotate at 310° and 20°   VPOINT Rotate at 310° and 10°   RIGHT

VPOINT Rotate at 310° and 40°   VPOINT Rotate at 310° and 60°   TOP

FARCAD	TITLE:		DRAWN BY:	DATE:	CHK:	REV:

ACTIVITY 16
Add the given dimensions in model space using viewport specific layers
or in paper space using DIMLFAC

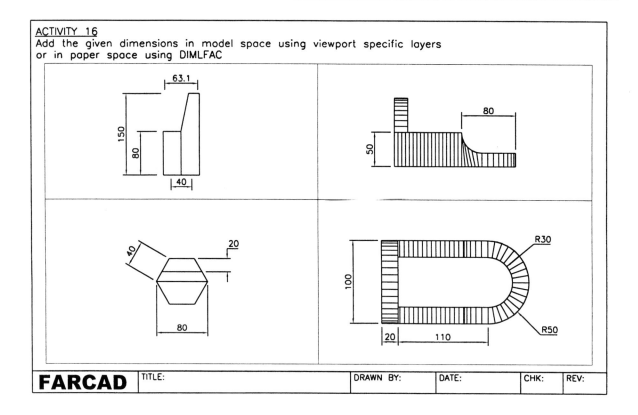

| FARCAD | TITLE: | | DRAWN BY: | DATE: | CHK: | REV: |

ACTIVITY 17: USING THE BASIC PRIMITIVES
Create a layout of your own design using the six primiitives.

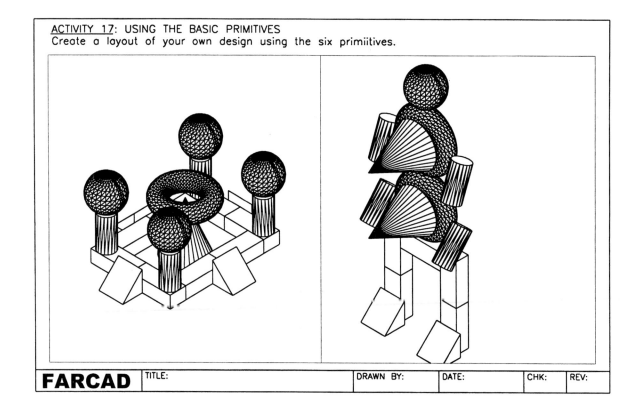

| FARCAD | TITLE: | | DRAWN BY: | DATE: | CHK: | REV: |

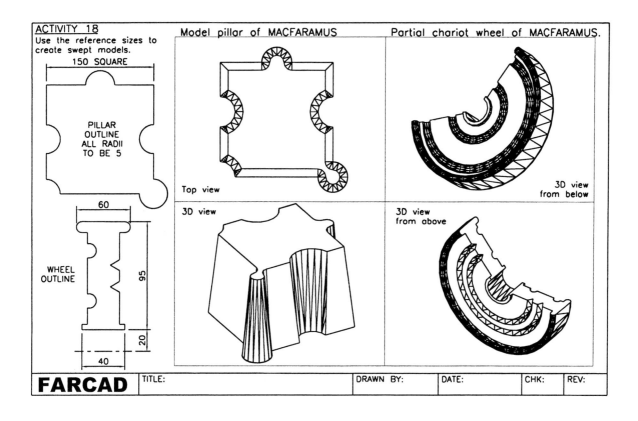

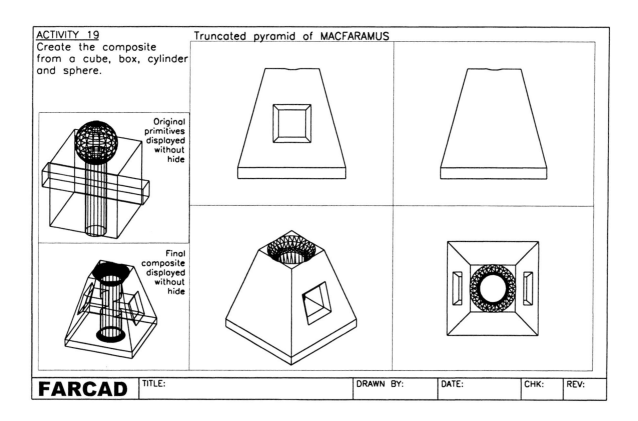

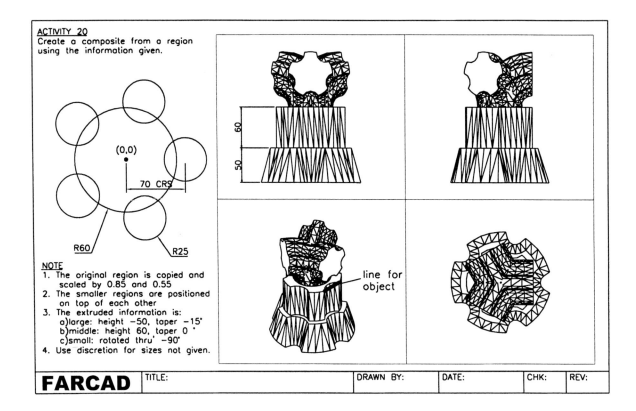

ACTIVITY 20
Create a composite from a region using the information given.

(0,0)

70 CRS

R60   R25

60
50

line for object

NOTE
1. The original region is copied and scaled by 0.85 and 0.55
2. The smaller regions are positioned on top of each other
3. The extruded information is:
   a) large: height −50, taper −15°
   b) middle: height 60, taper 0°
   c) small: rotated thru' −90°
4. Use discretion for sizes not given.

FARCAD | TITLE: | DRAWN BY: | DATE: | CHK: | REV:

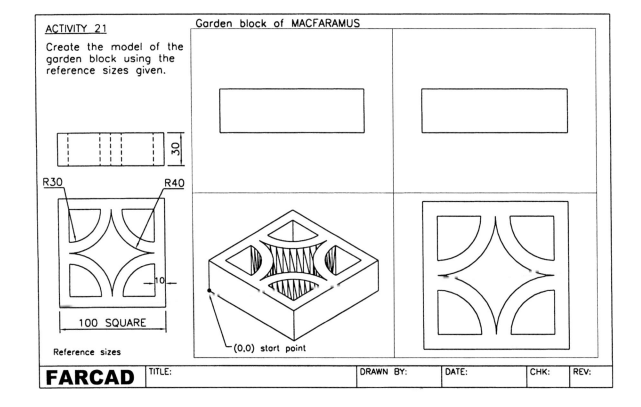

ACTIVITY 21

Create the model of the garden block using the reference sizes given.

Garden block of MACFARAMUS

30

R30   R40

10

100 SQUARE

Reference sizes

(0,0) start point

FARCAD | TITLE: | DRAWN BY: | DATE: | CHK: | REV:

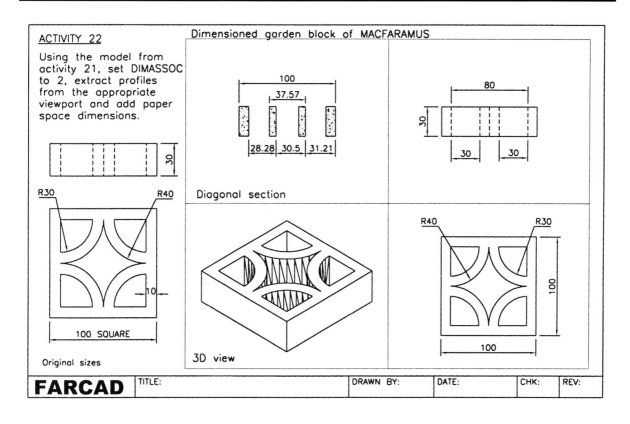

ACTIVITY 22

Using the model from activity 21, set DIMASSOC to 2, extract profiles from the appropriate viewport and add paper space dimensions.

Dimensioned garden block of MACFARAMUS

100
37.57
28.28 30.5 31.21
30

80
30
30 30

Diagonal section

3D view

R30 R40
10
100 SQUARE
Original sizes

R40 R30
100
100

FARCAD | TITLE: | DRAWN BY: | DATE: | CHK: | REV:

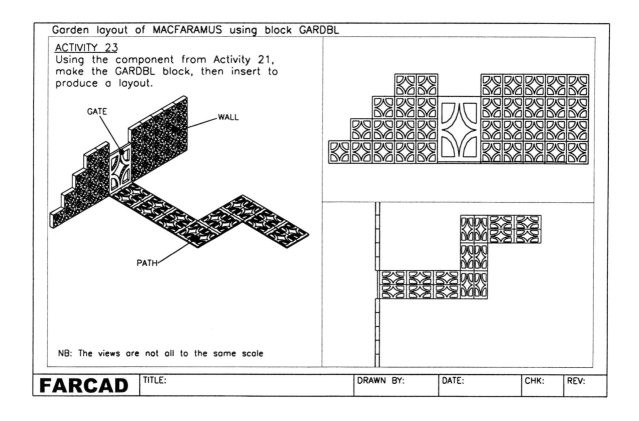

Garden layout of MACFARAMUS using block GARDBL

ACTIVITY 23

Using the component from Activity 21, make the GARDBL block, then insert to produce a layout.

GATE
WALL
PATH

NB: The views are not all to the same scale

FARCAD | TITLE: | DRAWN BY: | DATE: | CHK: | REV:

ACTIVITY 24: DISPENSER OF MACFARAMUS
Using the information in Fig 44.3 and
your discretion, create the views with
the SETUP commands.

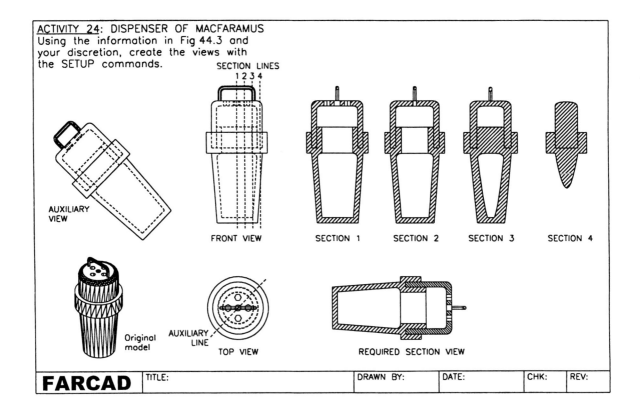

SECTION LINES
1 2 3 4

AUXILIARY
VIEW

FRONT VIEW   SECTION 1   SECTION 2   SECTION 3   SECTION 4

Original
model

AUXILIARY
LINE
TOP VIEW

REQUIRED SECTION VIEW

FARCAD	TITLE:		DRAWN BY:	DATE:	CHK:	REV:

Index